D0945207

# ANIMAL LANGUAGES

# ANIMAL LANGUAGES

## by Fernard Méry

translated by
Michael Ross

SAXON HOUSE

SAXON HOUSE
D. C. Heath Ltd.
Westmead, Farnborough, Hants, England

© Éditions France-Empire 1971
First published in French as
*Les Bêtes aussi ont leurs langages*
© Translation 1975 D. C. Heath Ltd.

ISBN 0 347 00040 1

Printed in Great Britain by
The Garden City Press Limited, Letchworth, Hertfordshire SG6 1JS

# CONTENTS

# CHAPTER I

# Introduction: animal languages

From the time men first studied the zoological world, it has been recognised that animals can communicate with each other, albeit by codes as yet little understood. This belief is confirmed daily by the most familiar animal sounds: the cock crowing at dawn, the nightingale singing in the evening, the mewing of cats, and the barking of dogs. Yet we still understand little about these forms of communication, their limitations and their development. Are they innate, stereotyped and exclusive to a particular species, or are they carefully acquired? Can they be adapted, perfected, even transmitted between different species? Does an animal select from a range of possible attitudes, gestures, cries or other signals the one which best conveys its "state of mind" or feelings at a given moment? Most people would not doubt that choice is involved. When an angry dog nearly chokes itself to death with its frantic barking, or shows its affection by yapping plaintively, or when it bares its teeth or wags its tail in impatience, surely it is expressing what it *wants* to express?

I

Here is the root of all the misunderstanding. Certainly, so far as dogs are concerned, we know what is implied by whining, barking, growling and romping. But does the dog know, and is it aware that it knows?

Over a period of at least 10,000 years, the dog has become man's servant and friend. The development from mutual mistrust to mutual tolerance has been so gradual that ties of great emotion have come to influence our hearts and minds. But because we have been so long content to attribute human qualities to dogs, we are just as far from understanding them as we are from understanding most other animals. Appropriately, J. Filloux has posed the question: "How long will it be, before animal lovers (basing their conclusions on thousands of examples and mistaken beliefs) cease to infer that because an animal's behaviour seems to resemble their own, the animal is experiencing similar feelings?"

Modern animal psychologists observe, note and make comparisons, but are careful not to draw conclusions outside the range of known facts – a very wise precaution, especially when we come to tackle such a vast and delicate subject as "language".

True language embraces the conscious use of signs and words in order to convey emotions or thoughts to others. Within the exact terms of this definition, animals have no language because they are unable to express abstract ideas. Apart from the much discussed "emotional consciousness", should we attribute awareness of *self* to an animal?

Observations and experiments are there to prove that animals do have this awareness. The behaviour of all animals depends on sensory phenomena which in many cases differ entirely from our own physiology. Analogies must therefore be discarded. If, on the whole, animal intelligence has remained a mystery to us, even in modern times, it is primarily because it is of a different order from ours. This is the simple fact which has taken us thousands of years to recognise.

Who would have guessed only fifty years ago that certain sense organs had such different uses, depending on the animal? Because their eyes are not purely for seeing, but are also sensitive to infra-red rays, certain predators are able to hunt in complete darkness. When, for example, a fieldmouse shelters at night in the hollow of a furrow, it radiates body warmth. The predator has no need to see its prey, but, teleguided by this animal warmth, it can pounce blindly on its victim. Again, without the discovery of supersonic waves and the knowledge that bats emit 40,000 of these waves per second, we could not understand how this living sonar, flying at nearly forty miles an hour, can pass between two black silk threads placed 8 inches apart, without touching them.

How could we imagine, barely twenty years ago, that a dolphin could adapt itself to the company of man, or foresee that in the ocean depths it could not only pinpoint an object, but determine its nature and even the direction and speed at which the object is moving? How could we know that at mating time a moth can locate a female miles away unless we knew that the female's body temperature is several degrees higher than its immediate environment and that this heat is irradiated according to a rhythm of variable frequencies and points of intensity on different wavelengths? This mystery was solved with the discovery that the male moth possesses antennae composed of fine "hairs" of constant length (or of multiples of this length) and that the variations of these proportions are the same as the frequency variants emitted by the female.

Similarly, Rémy Chauvin has proved that all kinds of communication can exist between several ant colonies. After introducing radioisotopes into various nests of red ants, it was clearly established that other nests, as much as 50 yards distant, also became radioactive: strangest of all, these colonies communicate with each other in an amazing way – their vibrations are transmitted through the ground.

It is not surprising, then, that animals can communicate with each other by means which for centuries were beyond our comprehension. Mistakes were made concerning the nature of these languages because we gave mere "signs" or gestures the value of "words". Whether these signals are made unconsciously rather than formulated is relatively unimportant, but the code which they form is unquestionably understood by all members of the same species. Thus the exact territorial boundaries of cats can be established by the smell of their urine and crows assembled in a field can be dispersed by loudspeakers reproducing their distress call. It is possible to make a duck and her ducklings take flight in a given direction by projecting on the ground the stylised outline of a certain predator. Professor Lorenz was able to make goslings obey him by imitating the honking of their mother. In this way, it was also learnt that seagulls, four days before they were hatched, could hear their mother's individual call through the eggshell.

Many other strange phenomena were unexplained until recently. As a final example consider the hen who, as soon as she hears her chick cry, will rush to its aid even if she cannot see it. But place this same chick under a transparent cloche, and the hen will remain indifferent to its visible but now inaudible distress.

Do we know any more about the origins of our own symbolic and articulated language and about the part played by this marvellous privilege in the physical and mental evolution of the human race? Some of the greatest contemporary authorities and research workers have devoted their attention to this problem. But once beyond the threshold of the brain's anatomy and physiology, scientists are just as baffled by the inadequacy of the gorilla as by the phenomenon of the human genius, and are obliged to fall back on words such as "chance" and "miracle". Any attempt to interpret the slender means whereby the closed animal world – from invertebrates to the most sophisticated of mammals – expresses its feelings, needs, desires and emotions must obviously involve making accurate distinctions between the various sensory, mental and social aptitudes of each species.

# CHAPTER 2
# Insect language

The "language" of insects remained a mystery until relatively recently. It was, in fact, the first to be decoded by science, before the "languages" of more familiar creatures which we thought we knew and understood. The most interesting discoveries have been made in the study of bees, wasps, termites, ants and butterflies.

In modern research on invertebrates, the name of Karl von Frisch immediately comes to mind. Before the First World War, this great Viennese zoologist was already widely known for his work on the colour vision of bees. Around 1949 he won world fame by demonstrating that bees – if not able to speak – had a most unexpected means of communicating amongst themselves.

Von Frisch placed a tub of flowering plants in a deserted spot. Some distance from here, a bee, selected at random, was taken from its hive and brought near the tub of flowers. The bee flew round this fragrant manna for a moment, made contact, gradually helped herself to nectar and pollen, then flew away to rejoin her sisters. On arrival at her hive she installed herself in their midst and mimed a "conversation" which informed them of her discovery, the place from which she had just come, the distance at which it lay, and the nature and quality of her treasure trove.

To begin with, von Frisch's discovery was regarded with amused scepticism, but he soon produced indisputable proof of his findings. No sooner had the "scout" finished her report than a commando task force of bees was organised and flew off to the flowers she had indicated. Now, this detachment set out alone; at no time did the bee scout guide them. Von Frisch demonstrated this conclusively with a transparent Perspex hive, through which he could observe the arrival of the discoverer, the mimed conversation which followed and the departure of the commando detachment. What was more, the bees had their own personal "language", a mode of expression consisting of gestures, perhaps the first form of communication between men: they danced!

Today, we know more about their bewildering performance. The bee's dance begins in the dark interior of the hive with a sort of fluttering movement on the wax honeycomb and a discharging of liquid secretions. Then, taking flight, the bee traces a figure "8", raising the tip of her abdomen and vibrating her wings. Von Frisch calls this the *Rundtanz* (or round dance). It becomes more rapid and lively according to how many flowers there are and how good the nectar is. Then, outside the hive, a sonorous and fluttery *Schwanzeltanz* (tail-wagging dance) follows to indicate the direction and distance to be covered.

Von Frisch experimented further.[1] By using arti-

8

ficial "bases" (sugared water in small cups), which he placed at a distance farther and farther away from the hive, he found that the more distant his experimental base, the slower the dance and the less precise the movements. But von Frisch wanted to be even more accurate and to calculate quantitatively the time spent on the dance in relation to the distance at which the "base" was situated. At distances between 100 metres and 10 kilometres from the hive, he noted that the rectilinear segment of the dance always took fifteen seconds (see plan). The greater the distance, however, the fewer the number of "waltz" revolutions in fifteen seconds.

This was undeniably a means of transmitting a message. In fact, by displacing one of the artificial "bases" 16 inches after it had been found by the bee scout, he observed that the commandos did not fly to the saucer containing the sugared water, but to the spot where the saucer had been and to which the bee scout had so accurately directed them.

How had she done this? The answer lay in the geometric formation of her dance in relation to the position of the sun. It did not matter that the sun was not always visible; in fact it might be obscured by clouds or completely hidden during bad weather. Bees know where the sun is at every hour of the day from polarised light which our poor eyes (without artificial aid) cannot see, any more than we can see ultra-violet rays.

Actually, three methods of communication have been established in bees:

1 *Touch*: By a pulsation of their antennae, two bees can identify each other. Their antennae

possess a receptive organ which allows them to control the scent of their Nassanoff's glands.

2  *Sense of smell*: Nassanoff's gland, common to all drones, is situated on the back of the bee and secretes a scent which the bee can release at will by merely bending her body. The scent is characteristic of the hive to which the bee belongs. Furthermore, it is sufficient for the explorer bee, already fed to repletion on floral nectar, to open her crop and release some of its contents for the whole hive to identify the flower.

3  *Gesticular mime*: We have seen the eloquence of the miming dance discovered so amazingly by von Frisch.

Ants present similar patterns of behaviour and are less hazardous to observe than bees. For example, after placing four saucers upside down on the ground, each containing a drop of honey, catch an ant some 20 yards away and mark it (as gently as possible) with a drop of plaster or white paint so as to recognise it later. Now pick up a saucer at random – the second one, for example. Place your ant under the saucer and let it feed for a few moments. Next, carry the ant on a twig and return it to the nest. In less than three minutes hundreds of anonymous ants will be seen marching directly to the second saucer.

Why? One might first assume that the ants are attracted by the smell of honey, but each of the four saucers contains honey. Or they might have been led back by the ant guide, easily recognisable because of the white marking. But she has remained in the ants' nest. Her sisters had no need of her to guide them directly to the second saucer and to that one only!

.  .

# The chemical language of olfactory signals

This is the way certain ant species behave, but not necessarily all, for myrmecologists (specialists who study ants) are continually finding new methods of communication between different species. According to whether the ants are African, Asiatic, European, etc. – and 6,000 different varieties have now been classified – each behaves in a distinct fashion.

Dealing specifically with ant scouts, Hingston[2] in 1928 drew attention to a distinct variant in the *Crematogaster auberti* ant, from the neighbourhood of Baghdad. When a scout finds fragments of an insect's carcase, it returns to the nest and – like other species – recounts its adventure. But it does not remain there. Excitedly jostling other members of the colony, it offers them a little of its nutritious find (with which it has already gorged itself), then accompanies a work party back to the scene of its discovery. Who selects this party and decides on the number of participants? And how is this done? We do not know, but the size of the work party always corresponds to the value and number of fragments of the prey.

Modern myrmecologists have tried to interpret this "language", but the behaviour of the *Cremato-gaster auberti* ant seems less rigid than that of other ants. The question is whether the *Crematogaster auberti* (and probably other varieties) possess a "symbolic" language adaptable to the circumstances. According to a number of American entomologists this is by no means impossible. Certain ant species can exchange information in a still more sophisticated manner than the *Crematogaster auberti*.[3] Discussions seem to take place between the informant ant and the rest. They form little groups, touch each other lightly with the tips of their antennae or legs, then separate again, repeating this until a final selection of "commandos" is ready to march forth. One gets the impression that the ants have come to a decision and that the "scout", having been questioned, has, by virtue of its superior knowledge, prevailed as leader. Obviously, there is no proof of this, and to regard the impatient or hesitant movements of the forager ants as the action of a discussion group is pure anthropomorphism. We must beware, therefore, of regarding what is probably purely an instinctive reaction as conscious reasoning.

Nevertheless, in the view of E. O. Wilson,[4] one of the greatest authorities on the subject in the United States, all ants can transmit different signals according to circumstances. Certain species, so he claims, are even able to emit signals at variable and successive intensities, just as we join words to form sentences.

What is the nature of these signals? To what exactly do they refer? On this point not all myrmecologists agree. It has been known for a long time that, because of their dark habitat, ants cannot rely on visual stimuli to any great extent, but transmit certain information by touch – by tactile pulsations of their antennae. Rémy Chauvin[5] has never been convinced – unlike Szlep and Jacoby – that these confused pulsations and trembling of antennae form a code (for example, to induce the forager ants to leave their nests).

Chauvin, the foremost French myrmecologist, attaches more importance to olfactory signals. This is a "chemical language" so rich that among the *Solenopsis sœvissima* (a species of "fire ant"), no fewer than seven different olfactory means of communication have been observed. There is the scent of the nest, scent for marking out a trail, scent for a summons to battle, and so on.

These odoriferous liquids – or pheromones – are secreted in glands whose anatomical position and physiological functions are becoming better known. Among other glands, as Rémy Chauvin tells us, is Dufour's gland from which a colourless secretion flows along the length of the ant's sting. This secretion is used to mark out a trail, the ant touching the ground with its sting at regular intervals, leaving a sort of dotted line. Other glands close by secrete liquids which are substances of alarm. Fewer than a dozen of these different pheromones are enough to control the whole organisation of an ant nest.

However, to explain the complex and various ways in which this "chemical language" is employed by different ant species – at least, those most familiar to us – one would have to make a complete résumé of the works of Rémy Chauvin.

## Wireless among the ants

Apart from these olfactory signals, ants have other quite different signals which a study of the remarkable *Solenopsis sœvissima* has revealed. The *Solenopsis* was first observed some forty years ago in Texas and immediately made its presence felt. Not only is its sting extremely painful, but, worse still, it is poisonous enough to kill a child.

It was obviously necessary to rid society of this pest but the most potent insecticides were useless against it. Even Dieldrin, the effects of which are both rapid and irreversible, failed to destroy the offspring although it massacred the fully grown ants. It was then that Wilson, digging deep into the soil, discovered that the females, after the first dusting with Dieldrin, immediately took evasive action by hurriedly removing all their eggs and burying them many feet underground. Thus, unexposed to the poison, the future ants were safe. But why did the female ants in neighbouring colonies, unaffected as yet by the poison, follow their example?

Intrigued, Wilson and his pupils set out to get at the truth. Just as von Frisch had studied bees, day after day Wilson and his companions now patiently devoted themselves to deciphering the code of these intelligent ants.

It had already been shown that, apart from olfactory signals, ants emitted sound signals on frequencies between 20 and 100 millihertz (which are inaudible to the human ear). When research was continued on these lines, it was soon discovered that these highly acute sounds, produced by the ants rubbing one segment of their abdomen against another, were transmitted by vibrations through the ground. Not only could these sonic emissions alert the whole ant nest but they could also warn other colonies, even those relatively far away.

These few illustrations reveal the danger of comparing the innate methods of communication among certain social insects, not only with the intelligent language of men, but even with means of communication between other insects which merit a study of their own.

For the sake of clarity, therefore, we can say that patient research has now yielded details of the sensitivity of various species of insects: termites and butterflies are mainly olfactory; dragonflies are visual; bees, both olfactory and visual; ants, tactile and olfactory; and finally, crickets are auditory in the extreme.[6] This relative classification will give the amateur entomologist a simple guide to the different ways in which insects emit signals and answers; in short, how they "communicate".

## Luminous and odoriferous creatures: "warning" languages

Many "signals" constitute what can conveniently be called "warning languages". These spell danger or threat and also indicate a presence or are used as calling signs. Some of these are luminous: for example, at the moment of mating glow-worms and fireflies "light up" instinctively, just like warning lights on the dashboard of a motor car. These "calling lamps" are not illuminated by chance, but are regulated by the insect. Fireflies, gathered together in great numbers on neighbouring trees, will all light up or become extinguished simultaneously, following some nervous rhythmic discharge. But how this occurs remains a mystery.

Among other invertebrates, the colour of the skin comes into play. In repose, for instance, the hawk-moth caterpillar (among others), assumes the shape of a twig covered with creamy-white lichen. But when a predator appears, the caterpillar quickly turns round exposing merely the front part of its body, decorated with a wide green band. At the same time the camouflage is perfected. The legs disappear, hidden away in the creases of the body segments, while on its stomach appear two large black spots, like eyes, hitherto invisible. Within seconds the timid caterpillar has taken on the likeness of a mini-serpent – a transformation that repels attack.

Nature has endowed her creatures with even stranger ways of deceiving enemies. Take, for example, the Hairstreak butterfly which sports on its wings two spots in the form of eyes, each surmounted by a fine filament resembling an antenna. For a long time it was believed (without definite proof) that predators, deceived by these false eyes and false antennae, attacked their victim at this very point; thus the butterfly would save its head.

This theory, however, lacked scientific support, and so Swynnerton decided to test its validity. With an extremely fine brush he painted similar false eyes, or ocelli, on the wings of fifty-one butterflies (genus Charaxes) then set them free. When he recovered them some hours later, the experiment proved conclusive. Although the butterflies had been pecked forty-one times, thirty-six of these attacks were precisely on the false ocelli. The birds had been fooled.

Other insects use odoriferous "signals" to ward off enemies. For instance, the golden scarab (a carnivorous beetle) gives off a disgusting smell like that of a crushed bug. Obviously, we do not know what attracts or repels the predators, but this particular smell must be one of the most effective since all the higher mammals, including man, react to it sharply.

It is easy to appreciate the difficulties which confront research workers trying to decipher the complicated codes of different insects, and, above all, to determine the exact anatomical centres from which insects emit their "messages".

Many years ago, Fabre devoted his attention to the mysteries of the mating calls and responses of different sorts of butterflies. Having placed a female silk-worm moth in a fine wire mesh cage, he noticed that in less than twelve hours more than fifty males, attracted to the female, had crowded round the cage. The same thing happened when he substituted a giant peacock butterfly for the silk-worm moth. On the other hand, when he placed one or the other of the two species in a sealed transparent cloche, not a single male appeared: unable to pick up any odour, they could receive no message.

More than a century later, Professor Adolf Butenandt added his contribution to our knowledge of the olfactory language of insects, in particular to that of butterflies. Butenandt succeeded in isolating this famous odoriferous substance which caused so much excitement in male butterflies. From 500,000 glands of the female silk-worm moth he obtained 6 milligrammes of this highly disturbing perfume. No sooner had he placed this substance in a laboratory dish, than clouds of males arrived from all sides.[7]

Such a discovery was bound to lead to another. In the same laboratory at the Max Planck Institute, Dr Dietrich Schneider attempted to prove that the method of communication which alerted males was not chemical, but rather a long-range mechanical shock. There was, he maintained, no reaction between the odoriferous substance and the components of the butterfly's nerve cell but a traumatic shock to this cell when pierced by the scent molecule.

Until we can manufacture these odoriferous substances synthetically for our own purposes (for example, the fertilisation of flowers by remote control) the results of such scientific research might seem a little like deciphering Egyptian hieroglyphics or the many other little-known languages of human civilisation.

The mystery of the butterfly's sense of smell was only partially unravelled. A few years later, Fernand Lot reported that, by pure chance, a distinguished French scientist had made a further breakthrough. One morning, Olivier, the four-and-a-half-year-old grandson of Professor Paul Portier (author of the *Biologie des lépidoptères*), had been greatly intrigued by something his grandfather had shown him. After placing a butterfly on its back with its wings outspread, he had simply rested his finger gently on its thorax and the butterfly fell asleep – or, more exactly, fell into a cataleptic trance. To revive it, he had merely to open a bottle of perfume, or even vinegar or mustard, close to the sensitive butterfly. How could this be explained? Obviously by its sense of smell, but which organ reacted to the smell?

Suspecting that it might be the antennae, the professor began to segment them carefully. When the butterfly was again put into a state of trance, it again awoke to the perfume. Next, Paul Portier thought that perhaps the legs held the secret. And so, encouraged by the child, he segmented the legs. Once more the butterfly's trance was broken by perfume.

It was then that the frustrated child had an idea as bizarre as it was cruel: "Grandpa", he said, "let's cut off its head! Then we shall see." When the old scientist had beheaded the butterfly he let a few drops of verbena fall on the table, and the butterfly once again seemed to regain life. Now there was nothing left but the body, and here lay the root of the mystery. The olfactory sense of butterflies lies in their thorax and abdomen in the form of tiny stigmas (part of the style or ovary surface that receives pollen in impregnation). Here was proof that it is only when butterflies settle and make direct contact with flowers that they can truly smell.

## The coloration of butterflies and its purpose

Despite Fabre's experiments and the mutilations perpetrated by Professor Portier it would be wrong to think that butterflies are sensitive only to scents and that their means of communication is purely olfactory. To persist in this error is to ignore the purpose of the colour and rich detail of butterfly wings.

If the scent of a garden tiger moth is enough to keep a fieldmouse at a distance, and if one can discern among butterflies such different scents as those resembling chocolate, lilies or urine (without being able to explain the reasons for such resemblances), it is reasonable to assume that the lovely colours of butterflies have an importance which goes beyond their aesthetic appeal to man. In most cases, in fact, the effect – if not the purpose – of these brilliant bands, ocelli and polychromatic designs is that the butterfly, feeble and unarmed as it is, can defend itself by upsetting its enemies.

At least this is the conclusion reached after intensive research by the English entomologist Miriam Rothschild.[8] As a result of her work, it has been established that many harmless butterflies whose markings are exactly the same as those of poisonous species would have disappeared long ago but for this inexplicable camouflage. However, this is to broach the unfathomable and mysterious question of mimicry which is somewhat outside our main subject – means of communication.

To show that butterflies are essentially visual creatures, Tinbergen studied the behaviour of the female grayling when pursued by the male of the species. He hoped to discover which stimuli generate pursuit once the call has been given by means of scent. To find the answer, Tinbergen used bait.

Georges Thinès[9] drew the following conclusions. The colour, size and form of the bait is unimportant. On the other hand, the luminosity, distance and movement of the bait proved to be very significant. For the male butterfly, the typical image of a female would appear to be a very dark coloured object, flying very close and erratically.

Many other mysteries have surrounded the sensitivity of butterflies. We know that, like ants, they react perfectly to ultrasonic waves, in the range emitted by their enemy, the bat – between 40 and 80 kilohertz. Scheller and Tim have proved this clearly by using a special transmitter. At a signal of a certain frequency, butterflies take rapidly to flight, then, as the intensity changes, they either speed up or fall to earth like dead leaves.

It has taken almost a century to demonstrate this discovery. As far back as 1890, John Lubbock, Director of the London Natural History Museum, was already talking of "sounds which we cannot hear". Unfortunately he made the mistake of adding "... thus insects can communicate their ideas and feelings among themselves". (Butterflies with feelings?)

## Hello! Hello! ... crickets speaking ... over to you cicadas

Even without making comparisons and analogies, the most original scientific research sometimes seems amusing.

Today it is a well known fact that crickets have noisy and indiscreet love-affairs. Rémy Chauvin[10] recounts how Professor Regen, after observing that the female cricket, when summoned by the male, will make her way towards him from a distance of several yards, conceived the idea of recording the male's strident call on a microphone and then transmitting the sound by loudspeaker from a neighbouring room. This was sufficient: the amorous cricket unhesitatingly made straight for the loudspeaker.

Since then, whole orchestras of insects such as cicadas and crickets have been meticulously recorded by means of ultra-sensitive recording apparatus. By varying the speed on the receiver it has been possible to establish that the participants (in their capacity of receiver/transmitter) emit their respective sound signals in a curious way. The one that "attacks" first conducts the song, while the other accompanies it. If the leader stops suddenly, the accompanist will also immediately fall silent. But should his accompanist stop, the leader will continue "playing" in complete disregard of the other's silence.

Rather than discuss the reasons for this ourselves, we will once again turn to Rémy Chauvin to appreciate the complexity of this vocabulary which has successfully been put on tape, but which so far no one can decipher. "First of all, many insects", writes Rémy Chauvin, "are capable of not one, but several stridulatory rhythms [i.e., shrill, jarring sounds]. The

21

basic element is the utterance of one or more groups of vibrations of a definite frequency, comparable to the phonemes [i.e., units of significant sound] of human speech; a phoneme is separated from its successor by a definite interval, varying according to species, and the cadence [i.e., fall of voice or measured movement] of successive phonemes emitted is the first element in recognition. But other variations can be produced and the cricket's song is a good example: his courting song differs from his call song in that each train of impulses is less intense and more 'basic', that is to say he starts and ends less abruptly; each trill of song contains twice as many impulses; finally, between two trills, the male gives a short and sharp sound quite separate from the others."

All sorts of combinations are possible according to the frequency and cadence of the sequence of impulses, the length of the phrase in relation to the silent interval which separates it from the following phrase, and the regularity or irregularity of this interval.

"Nevertheless", adds Rémy Chauvin, "we can find within each song a 'grouping of groups' which further adds to the complexity. Emissions of a given frequency, in groups separated by a certain interval, may form part of a 'phrase' of a given length and be separated from the following phrase by a much longer silence."

Among insects – especially social insects – the means of imparting information and of communicating can thus be regarded as a kind of "language", a word which von Frisch was not afraid to use. He often uses the word *Bienensprache* ("bee-talk") when referring to bees.

According to Professor Piéron: "The main evolutionary difference between invertebrates and vertebrates is clear: the first (with bees and ants reaching the highest state of evolution) have concentrated on the development of instinct; in the second (with man at the top of the scale) it is intelligence that has been developed."

These words bear thinking about. For who can be certain that the evolution of these two worlds will be maintained, and that, in our case, having reached a point of sublime perfection with man, it will not tip over into madness? We know absolutely nothing of the mutations which gave rise to "intelligence" and even less of the origins of instinct and language.

If we are to restrict ourselves solely to the field of language, it would be wrong to claim a correspondence between a species' position on the scale of creation and the psycho-sensory possibilities of communication between its members or with the rest of the world.

Despite their rank in the animal hierarchy, we shall see that for the most part, fish bear no comparison with the insect world. Fish live in a different world which we are only just beginning to explore.

## Notes

[1] Frisch, Karl von,  *Bees: their vision, chemical senses and language*, Oxford Univ. Press (Oxford, 1951)

[2] Thinès, G.  *Psychologie des Animaux*, Dussart (Brussels, 1966)

[3] Chen, S. G.  *Social Modification of the Activity of Ants in Nest-building* (1939)

[4] Wilson, E. O.  "Pheromones" *Scientific American* CCVIII (May, 1933)

[5] Chauvin, R.  *World of Ants*, Gollancz (London, 1970)

6 Filloux, J.-C.    *Psychologie des Animaux,* P.U.F. (Paris, 1956)
7 Dröscher, V. B.    *Mysterious Senses,* Hodder and Stoughton (London, 1964)
8 Rothschild, M.    "Les papillons qui se déguisent", *Science et Vie* (May 1970)
9 Thinès, G.    op. cit.
10 Chauvin, R.    *Animal Societies,* Gollancz (London, 1968)

# CHAPTER 3
# The watery world

Water-dwellers can communicate perfectly with each other – and not only fish but also the lower species, from starfish to lobsters. One has merely to watch the expressive pantomime of the fiddler crab on the beach during the mating season. This little crab, with his two unequally sized pincers, strategically posts himself where a female will eventually cross his path. As soon as he sees the female, he raises his longer "arm" and waves it about comically to attract the beauty's attention. If she continues on her way, indifferent to his appeal, he hurries to overtake her, once more taking up his position and repeating his performance exactly like someone on the roadside trying to hitch a lift in a passing car.

Fish, of course, cannot gesticulate in this way. But this does not mean that they are unable to express themselves.

For a long time it was thought that fish were deaf and dumb and had very little sensory perception. Yet nowadays the "lowest form of vertebrates" (as they are catalogued) are known to be better equipped than insects. They have a skull, a brain, bone marrow and several receptive sense-organs, in addition to the lateral line which is as vital to them as their inner ear and swim bladder.

However, in a broad sense they are of course very primitive creatures. Fish do not possess what we would call individuality. Their general behaviour rarely changes and their ability to learn is extremely restricted. Nevertheless, fish can transmit and answer signals. More and more is being discovered about their sensitivity to various kinds of stimuli.

Legend, history and scientific research all have something to say on this subject.


## Legend

If one is to believe the *Odyssey*, Ulysses was not the first navigator to listen to the siren song of mermaids. It is only natural that the long inaccessible three-dimensional world of rivers and oceans should have stimulated the imagination and given rise to fables. Dugongs and manatees are still mistaken for mermaids; we are always willing to believe in sea-serpents; and in our thirst for the marvellous we came near to believing that after a few months the dolphins of Saint Thomas' pool in California would be speaking American.

The great silence of the ocean depths is nothing but a myth. It is amazing that, with all the bathy-scopes, microphones and cameras of Commander Jacques Cousteau, Marie Poland-Fish, Dr Lilly and other famous oceanographers, Neptune's great king-dom was not penetrated sooner. If only science had been encouraged all over the world (as it should have been), we would long ago have forced open the doors of this marine kingdom. As it is, we are only just beginning to understand it. Is it surprising that mariners' tales, told in all sincerity, were regarded as wildly extravagant?

## History

We do not have to go back as far as the Flood. To the best of our knowledge, one of the first tales of noises made by fish dates from February 1803. The great German naturalist Humboldt[1] relates: "That evening as night was falling, the entire crew of the ship, one by one, was struck dumb with astonishment. My turn soon came. In the midst of the ocean, in calm weather, we suddenly heard an extraordinary rumbling noise like a roll of drums approaching closer and closer. To begin with we took it to be the sound of breakers, but very quickly the noise spread throughout all quarters of the ship and was particularly loud in the stern. It was a bubbling noise resembling air escaping from a boiling liquid. For a moment we feared that the ship had been holed, but soon the noise spread everywhere and we became aware of it from every part of the vessel. This strange phenomenon lasted about two hours; then all was silence."

## Scientific research

Today science explains what could not then be explained. This tumultuous rumbling which so scared Humboldt's sailors was most probably due to

27

the huge quivering movement of a shoal of maigres in the process of mating. Was this the sound of love-songs, long-sustained choruses of hundreds of thousands of bass voices? Not at all. The appropriate word for it here is "ventriloqual" because these drawn-out sounds, vibrating through the cavities of the swim bladder, depend on abdominal muscles.

But who at that time could have identified them accurately or predicted this other no less historic (and completely verifiable) event. The day after the destruction of the American fleet at Pearl Harbour, the United States, stunned by the suddenness and the extent of the disaster, anxiously anticipated further surprise attacks on naval targets. The U.S. Naval department lost no time in installing highly sophisti-cated warning systems in the most vulnerable bases. They had not long to wait. A few days later, at dawn, the alert was given and the worst was ex-pected; but nothing happened. Rigorous enquiries made into the cause of this false alarm revealed that the sound picked up by the warning system had been made by myriads of passing shrimps.

Since then, scientific research has confirmed the fact that fish produce the most diverse sounds, from drum rolls to the crackling of a forest fire, not to mention the clucking of hens, the whistle of a boiler and the whining of dogs. Dr J. B. Horsey has detected these sounds at a depth of some 700 fathoms. Today, all scientists agree that fish locate and identify sounds through their gills and also by numerous "receivers" situated on the lateral line running along their sides.

From then on, science has made great strides in this field of research, and we know that: "since sound carries four times as quickly through water as through the atmosphere, sounds are deflected from the surface back into the water rather than out into the air. Above the surface, therefore, nothing is heard."[2]

This physical phenomenon has long been known empirically by the native fishermen of Ghana. To detect possible shoals of fish close to their boats, the Ghanaians thrust a kind of enormous wooden spoon deep into the water then put the tip of the handle to their ear. By slowly pivoting this spatula they can detect the sound reactions of near-by fish and locate them.

Today there is a whole range of sound recordings of the yapping of octopuses, the low murmur of drumfish and the cries of pain made by wounded whales.

It is even possible to differentiate between the various sounds produced by the swim bladder, which acts as a sounding board (75–300 hertz), those made by the grinding of teeth in front of the pharynx and the stridulations of thirty different types of shrimps, including the curious crack of the pistol-shrimp.

We also know that fish are no more deaf than dumb. As early as the first century B.C., the consul Marcus Crassus knew perfectly well that he had only to clap his hands as he approached his fish pond to see his favourite muraena eel come swimming towards him.

Since those days, we have learnt that fish react differently according to the size of the shoal and according to the space they occupy.

Experiments have shown that fish are affected more or less by all kinds of stimuli: visual, aural, olfactory, chemical and electrical. The following are a few random examples of specific reactions to common stimuli:

## Visual

Gudgeons can express, by means of their eyes, mouth and gill coverings, a whole range of threatening gestures.

Flying fish cannot bear the shock of light rays hitting them at night.

Mackerel are sensitive to quite remote infra-red rays.

Siphonias possess the curious ability to change colour immediately they sense danger.

## Aural

Fresh-water Ostariophysi (a genus comprising among others, the carp, roach and bream) can make and perceive high-pitched noises, whereas cod make hardly any noise except for low, deep sounds (barely 50 hertz) when their enemy, the conger-eel, approaches.

In addition, Dr Arthur Myrboy has demonstrated that various species of fish are able to locate one another's positions by the use of sound signals. Having noted that sharks are attracted by the sound signals of damsel fish (their usual diet) Dr Myrboy transmitted a series of low frequency sounds (less than 800 hertz) by means of underwater loudspeakers. These sounds were exact reproductions of those made by damsel fish. Within a few seconds dozens of sharks began to arrive. The experiment was repeated several days running, always with the same results.

## Olfactory

Evans discovered that lysergic diethylamin acid will

make the most timid gudgeon aggressive, while benactizin will make the most pugnacious fish inoffensive. Teichmann asserts that an eel can differentiate perfectly between pure water and a solution equivalent to a single milligramme of phenylethylic alcohol in an area of water, 17 kilometres by 10 kilometres and 100 metres deep.

## Chemical

Von Frisch discovered a strange alarm substance affecting minnows. One hundredth of a square millimetre of skin from an injured minnow, added to 15 litres of water, is enough to terrify all the other minnows in the aquarium, making them plunge to the bottom. On the other hand, Pfeiffer, experimenting with some fifty salt-water fish, failed to detect this substance which seems peculiar to fresh-water fish.

## Physics

Pardi and Papi have shown that lalitre crabs, placed on any firm ground, will automatically make towards the sea, guided, as are bees, by polarised light.

## Electrical

Electrogenic organs (unidentified as yet in creatures other than fish) have been discovered in half a dozen species, including the electric ray, the electrophorus, the gymnotus (American electric eel) and electric rays.

The American physiologist, Frank Mandriota, has made a study of these electrical currents, particularly those of the African catfish, which exceed 200 volts. The electrophorus gives off more than 500 volts but in other species the current is much weaker. Obviously it is extremely difficult to study this phenomenon when the fish are in their natural state. It would indeed be interesting to know how some fish have acquired the unique ability to become living electric batteries, whose perfection baffles even the best electrical engineers.

Some specialists favour the theory that these electrical discharges act as sonar guides. Others prefer to regard them as assembly signals. All discharges of this sort can in fact be picked up by members of the same species at far greater distances than is possible with other social signals, even aural or visual ones.[3]

Be that as it may, there is a natural tendency for all marine creatures to form into groups and to act interdependently. Nearly all fish, big or small, gather together to search for food, to protect or defend themselves, to reproduce or to migrate. Fish seldom live independently.

We shall see the importance of these social stimuli, to which members of a group respond automatically, simultaneously and in unison.


## Open and close ranks

Few social creatures, during their life-span, provide a more striking example of communal living than fish. In the animal world there is scarcely anything comparable to shoals of herrings, sardines or mackerel. They are so dense, extensive and united that one might almost mistake these aggregations of individual creatures for a single organism made up of an infinite number of exactly similar parts. An extraordinary feature of these moving shoals is that they often assume the elongated elliptical form of some enormous and unearthly fish.

Many writers claim that during the night these almost constant formations split up, but, at the first light of dawn, once more fall into their former precise "ranks". From then onwards the shoal will behave in such an interdependent manner that one almost forgets that it is composed of individual fish.

Whatever the nature of the shoal, its members, male and female alike, always resemble one another physically to such an extent that tunny fishermen, after their first catch, know whether their haul will consist of small, medium or large fish. In any shoal, all the fish space themselves out equally according to their size and species. The classic illustration of this is to isolate two fish accustomed to living in shoals. It will be found that they immediately place themselves parallel to one another.

The search for food is a communal task. It is the sight of their prey as much as the noise made by the first fish in the act of feeding which releases a simultaneous alimentary reflex in all the others.

The mating season is always the same for members of the same species, and reproduction takes place in groups, rather than in isolated families.

Whenever a shoal is threatened by serious danger, the reaction is unanimous. Without a leader or any sort of hierarchy, the whole shoal will immediately adopt defensive tactics. All the fish will zigzag simultaneously, jumping and leaping to precisely the same height and at exactly the same time to within one hundredth of a second. How are these mass movements started by all and for all?

Pavlov provided several illustrations of such

behaviour, but the strangest is the one given by Allen concerning a shoal of sardines when attacked by a seamew: "The fish formed a very compact spherical mass about 1.50 metres in diameter, which moved along slowly at a depth of 3 metres (it would appear that the cohesion of such masses is maintained because each fish tries to keep away from the outside, where there is always danger, and to seek safety in the centre). Each time the bird plunged, the sphere opened up at the point of attack. After "breaking ranks" the shoal immediately re-formed to make another almost perfect sphere. The bird dived and attacked several times before coming up for air. On one occasion it attacked the shoal from below. The mass then immediately formed into a ring through which the bird passed without capturing a single sardine!"

Apart from the physio-psychological problems posed by these astonishing mass reactions, they show that if fish are not intelligent, they can at least instinctively adapt most expediently to circumstances.

### Communication with man?

Leaving aside the great marine mammals, which are the subject of separate research, numerous laboratory experiments concerning the possibility of training and taming fish have revealed that communication between fish and man is limited to conditioned reflexes. In Florida, a half-caste named Raoul has earned himself a reputation as an amusing entertainer. At Kay-West, near Los Angeles (Boulevard Roosevelt to be precise), Raoul keeps some sixty different species of sea fish which he has succeeded in training to answer to their names. You can choose to see Jopo the merou swim towards him when the order is given, or Ramona, the dory. Such demonstrations are, however, of the order of a circus act.

The research conducted by Professor Frolov (who enjoys the same reputation in America as a Jean Rostand in France) is somewhat more sophisticated. In a small aquarium, Frolov immersed a telephone receiver, which, at the touch of a button, emitted a slight buzzing noise. Every fish in the aquarium was attached to the receiver by an electric filament. Sufficiently fine and supple not to interfere with its movements, this filament transmitted a slight electric shock as soon as the button was pressed. The fish then darted towards the receiver with a movement of slight nervous impatience. After some fifty experiments, Frolov removed the filament and merely sounded the buzzing signal, but the fish still reacted in the same manner. Next he took the receiver from the water, but again the fish continued to jump slightly each time the telephone buzzed, although the receiver was now at quite a distance from the aquarium.

The question of intelligent communication between man and dolphin, however, is a different matter, and one which in the last few years has received world-wide attention from scientists. In Giraudoux's play, *Ondine*, it may be recalled, the poetic fairy creature brought up at the bottom of the sea sighs confidentially, "I know all about the little fishes". Today it would seem that we, too, know much about dolphins and their language.

But where do these studies enter the realms of science fiction and what light has been shed on this tricky problem?

# Notes

1 Humboldt, F. H. A. von, *Cosmos: a sketch of a physical description of the Universe*, H. G. Bohn (London, 1848–58)

2 Bauchot, M.-L. and R. *La Vie des Poissons*, Stock (1967)

3 Bauchot, M.-L. and R. op. cit.

# CHAPTER 4

# The lords of the sea

We come now to other creatures, which, despite their general appearance and their habitat, are no longer fish.

There are over a hundred species of cetaceans (marine mammals), of which practically nothing was known at the close of the last century. These giants of the sea vary in length from 18 to 90 feet and weigh from 400 lbs to 130 tons. They have lungs with which to breathe, warm blood flows in their veins, and their young are suckled. If we add that cetaceans have the largest brains in the world (the dolphin's weighs some $4\frac{1}{2}$ lbs; the common rorqual's more than 20 lbs) and that despite their bulk, most of them slip through the depths of the ocean as easily as birds cleave the skies, it will be appreciated how difficult it is to observe these mammals in their wild state, under conditions favourable to the study of their behaviour and mental processes.

Past civilisations have always been attracted by the wealth of mysteries surrounding them. For instance, it was said that Apollo took the form of a dolphin to turn the marvelling Cretan sailors off their course and make them the faithful guardians of his temple. Orion, when cast into the sea from a boat, was saved from drowning at the last moment by dolphins which escorted him safely to the shore. Jonah's fate in the belly of a whale was even more miraculous. All these classical legends point to the respect in which the Ancients held these "giant fish", which they sensed were different from the others.

Since then, right up to modern times, legends and anecdotes have been confined to a few dramatic stories of sailors "swallowed up by a sea monster". What should we make of these stories? It would be physically impossible for whales with baleen plates (i.e., the brush-like whalebones that act as filters in the mouths of certain species such as white whales) but quite possible for those cetaceans which have teeth, particularly the large sperm whale.

For the last twenty years information on marine mammals has been as inconclusive as it has been rare, and more so when attempts have been made to analyse the possibilities of "language" among the different species. Only by occasional good luck has any light been shed on the subject. In normal circumstances, as Commander Jacques Cousteau has observed, the great sperm whales allow their young to wander some distance (up to one or two miles) from the main school. Yet as soon as danger threatens, all the whales will inexplicably reassemble, somewhat distraught, in less than three minutes.

On one occasion, the crew of the research ship *Calypso* was fortunate enough to witness such a scene and managed to film it. First, a huge male raised itself upright and began to turn around slowly on its

38

own axis, like a radar scanner. Almost immediately the young whales rushed from all sides to join him. Cousteau's crew managed to immerse several hydrophones to the right depth and even to film this phenomenon. The giant sperm whale, weighing 100 tons or so, was dancing like a bee, for it was not only transmitting alarm signals but was also clearly receiving answers. We have learnt since then that, should the need arise, that is, at the slightest danger, one of the group will be delegated to set off immediately to bring back a youngster which has rashly wandered too far.

Some time later oceanographers on the *Atlantis*, equipped with the most modern apparatus, recorded more or less similar scenes and managed to identify three separate kinds of noises. The first, ill-defined and rather confused because the cetaceans were under water, resembled the sounds of muffled hammer-blows repeated every half-second. The next was like the creaking of a door on rusty hinges. Finally, the third sound (or, more precisely, series of sounds) consisted of a score of short clicking noises at the rate of five per second, each clear enough to be registered on the equipment's recording tape.

If these lords of the sea do have their own language, what form does it take, and what do the signals mean? The only way to find out was to capture a sample group of a suitably-sized species and keep them alive in huge salt-water tanks. By attempting to tame them, we would learn more about them, and perhaps even train them.

### After pet dogs, pet dolphins?

"Within the next decade or two the human species

39

will establish communication with another species: non-human, alien, possibly extra-territorial, more probably marine; but definitely highly intelligent."

It is on this optimistic note that Dr John Lilly opens his book[1] on his work with cetaceans, and dolphins in particular.

Of all the known marine mammals, it was quickly discovered that dolphins, and in particular the so-called "bottle-nosed dolphin" (*tursiops truncatus*), would be ideal subjects for the study of interspecies communication. In the first place, because of their character, dolphins have always been regarded as friends. What can account for the mysterious instinct which allows a dolphin to approach human beings without a trace of fear? We do not know, any more than we know why these cetaceans, whose terrible teeth are capable of cutting the most ferocious shark in two, have never attacked a man. And what remote sympathy makes them join in our games with all the happy confidence of a pet dog?

It was, however, for much more mundane reasons that Dr Lilly and other scientists chose the friendly dolphin as a suitable subject. A study of the language of marine mammals could only take place in captivity, using a species whose brain resembled ours in both size and complexity.

Many attempts had already been made to communicate verbally with certain higher mammals such as primates, but so far all experiments had ended in dismal failure. Why was this? Perhaps because the brain of even the largest anthropoid ape is four times smaller than that of man. (The average weight of a man's brain is about 4 lbs, while that of a chimpanzee is less than 1 lb.) The brain of a dolphin, therefore, was almost perfect for the study in hand because it weighs about $4\frac{1}{2}$ lbs. It is also more complex than man's – in folds, divisions and the number of cells in the cortex.

This, however, was not the only necessity. The selected animal had to be capable of producing by means of its vocal organs (either underwater or elsewhere) sounds like those of humans, or, at least, sounds which could be easily modified to fall within man's acoustic range. These criteria, of course, concerned only sounds, and did not take into account other means of expression such as smell, touch and sight. It was also necessary to select a relatively small cetacean so that the experimenter could handle all operations and control them without running any serious risks.

Only the dolphin fulfilled all these requirements, and it was therefore decided on.

Hardly had work begun when unexpected observations were made. The bottle-nosed dolphin was found to be quite capable of making sounds in the air. When it reared itself up out of the water, those within a yard or so could hear a gentle whistling noise issuing from the blow-hole near its "forehead" through which dolphins breathe. Similarly, it was soon found that dolphins could be trained and that their willingness to learn and obey depended greatly on the research worker's ingenuity and also on a system of rewards rather than punishments.

## The amazing dolphins and killer whales

We will not go into the technicalities of the capture, transport, accommodation, anaesthesia and resuscitation of these unusual "patients". The first to be successfully anaesthetised uttered its first distress signal as soon as it came round. Then two other dolphins, sharing the pool, replied to the call in their own way. Taking it in turns to glide under its head they applied themselves to the task of keeping it above the surface.

One must be careful not to read into these words any hint of anthropomorphism. As John Lilly went on to show, this was indeed an S.O.S., a coded sound signal. Whenever the dolphin, suffering from the effects of the anaesthetic, gave a whistle, its two fellow dolphins answered with a particular twittering noise and redoubled their efforts.

Two years later, Dr Lillys team had succeeded in perfecting the main methods of observation, enquiries into the nervous system and experimental phonetic studies. The publication of his pioneer tests and the foundation of the first Communication Research Institute soon followed.

During the next few years, the work continued with alternate success and failure. Each new discovery filled the team with enthusiasm, and every failure was a challenge to be met with renewed effort. Dr Lilly's reports aroused world-wide interest, and were given wide publicity by press and television.

We seek and always will seek novelty, and now the laboratory doors were beginning to open on a subject which bordered on the incredible and the miraculous. Henceforth, all marine mammals would be regarded as the most intelligent, and the most perfectly adjusted, of all the innumerable animals of earth, sea or sky. It does seem, however, that all cetaceans of the same species can communicate with each other in disturbing situations.

During an important conference at Dr Per

Scholander's Zoophysiological Institute an expert from the Whaling Institute remarked that the killer whale is also highly intelligent and perfectly capable of formulating a language. He related the following story in support of this statement.[2] During the previous Antarctic whaling campaign, a group of several thousand killer whales appeared in the area where a fishing fleet was operating. By killing the fish near these boats the whales made the fishermen's task impossible. In response to a radio call for help, the whalers sent several small whaling boats or *shytters*, one of which fired a single shot from a harpoon gun. Within half an hour the killer whales had completely disappeared but only from the vicinity of the *shytter* boats. Over a range of some 80 square miles there was not a whale to be seen, yet the fishing boats that were some distance from the *shytter* craft were still plagued by the whales.

The remarkable part of this story is that both the fishing vessels and the little whaling boats were converted World War II Corvettes. They therefore looked exactly the same, the only difference being the harpoon guns on the *shytter* boats.

In the narrator's opinion this demonstrated that killer whales definitely do use a code, a means of rapid oral communication, which is capable of both describing objects and indicating required action. Everything seems to indicate that in less than half an hour the killer whales had been able to give their companions a description of the dangerous *shytters* and thus to change their behaviour not merely for the next few moments, but for several hours afterwards. It had previously been thought that such a complex system of communication could exist only in humans.

43

Obviously this behaviour differs radically from that of a shoal of fish which suddenly changes direction in response to some mysterious signal. The school of killer whales went further, sending out a description of a dangerous object: a harpoon gun. "We can imagine", says Lilly, "the minimum amount of information that must have been transmitted and place it in sharp contrast to the kind of information a school of fish might send out."

It seemed as if a wounded killer whale had warned his immediate companions of the danger. They passed the message on to the other whales. ("Watch out for something which sticks out at the front of the boat! Something which shoots out a pointed thing which sticks into our bodies! Watch out! There is a long line attached to the end and they pull us towards their boats!") And then what happens? The whales glance towards the boats and continue their plundering only around the boats that are not armed with this strange "something".

It seems fantastic. The behaviour of ordinary fish would obviously have been totally different under similar circumstances. In human language, the leader of the shoal would have simply changed course ("Right turn"). In a split second, the order would have been given and acted upon. In the case of the thousands of whales the communication and adoption of a whole series of responses must have been involved ("Keep away from the dangerous boats. Don't worry about the others which look like them ...!").

Such behaviour is obviously too complex to depend purely on instinct, but to talk of a human brain in a dolphin's brain is to risk falling into the trap of anthropomorphic comparisons. He continues: "Who can claim or deny that cetaceans, gifted with a large and complex brain, do not engage in oral conversation although they are indeed incapable of talking written language."

To return to dolphins, it can be said that we have here a firm basis for research into communication between cetaceans.

## Dolphins talk to each other

It is within this specialised field that Dr Lilly's controlled experiments have contributed the most. Yes, dolphins do talk to each other. At least they seem to communicate, using sounds which we have so far been unable to decipher fully. The following account, though it is so well known, deserves to be given in full.

In his laboratory on Saint Thomas Island, John Lilly recorded the sharp whistles made by a pair of dolphins kept in the same pool, but separated by an opaque metal screen. Recognising each other's voices, the male and female called and answered one another. At first they exhausted themselves by leaping several times in a vain attempt to see each other over the partition, then they seemed to tire of this and gave up the idea. Silence prevailed once more and they stayed motionless at the bottom of the pool.

After some moments the male surfaced and then plunged to the bottom, attempting to give a call. He made whining noises and emitted little grunts of varying intensity. The female then replied, and the male immediately fell silent allowing his companion to chat away. For the space of half an hour there was an exchange of alternate vocal noises, each dolphin in turn waiting for the other to fall silent before resuming. It suddenly seemed to Lilly like a duet, "a sort of romantic song which in spite of the strangeness of the noises, seemed to me quite lyrical."

Subsequently, what was clearly a dialogue (although incomprehensible) was put to good use by many other animal psychologists.[3] These initial experiments gave Dr Kenneth S. Norris, who was engaged in research in the Hawaian Islands, the idea of putting the dolphins he was studying in the Pacific in telephonic communication with the dolphins housed in the aquariums of Florida on the Atlantic. "The speakers very quickly learnt how to use the underwater microphone and receiver (which were connected by radio link). Again each dolphin allowed the other to finish before answering. It was therefore obvious that dolphins speak the same language in all seas, which is not at all surprising since our seas and oceans are interconnected and have no frontiers."

Professor Winthrop N. Kellogg of the University of Florida went even further. While recording the frolics of a large number of amorous dolphins on the high seas, he showed that a pair could communicate perfectly well with each other, even if some distance apart and even when surrounded by a noisy group of other dolphins. This phenomenon (found also among colonies of penguins) is now known scientifically as the "cocktail effect", an allusion to the human ability to follow a normal conversation when surrounded by the hubbub and gossip of a cocktail party.

The theory that the dolphins' alternating patterns of sound constitute an organised dialogue was further strengthened by an experiment carried out on five porpoises, members of the great dolphin family, by Dr John Drener, Dr William Evans and Dr J. H. Prescott of the Lockheed Society, California.[4] While a number of these cetaceans were frolicking 300 miles off San Diego, the entrance to a lagoon used by the porpoises was blocked by a chain of buoys. Towards the late afternoon some porpoises were seen returning to the lagoon. On being confronted by these unusual obstacles, they hesitated, turned back and assembled in the shallow off-shore waters. Presently a "scout" left the others and went to examine one buoy after another. On its return, its companions greeted it with a storm of whistles. Then they stopped. Next, a second porpoise set out to investigate this puzzling barrier, being greeted on its return with another chorus of rapid whistles which quickly calmed down. Then the porpoises, silent but reassured, slowly dived towards the lagoon.

Two or three years later, American engineers from the Marine Experimental Station at Pasadena (California) continued Dr Lilly's experiments on communication between pairs of porpoises, by placing each one in a separate tank linked by an underwater telephone. The difference between Dr Lilly's experiments and these was that telephonic communication could be interrupted or reconnected as the experimenters wished. The transmissions were recorded on tape and each porpoise was given its own track.

In a flash[5] the two porpoises realised whether or not they could communicate with each other – just as if it had been a telephone conversation between humans. To begin with they always gave the same whistle which might be interpreted as the call, "Hello! Hello!" Another represented identification:

47

"Dash here" or, alternatively, "Doris speaking". Between these two very distinctive signals, a whole series of quite unexpected and wide-ranging sounds were recorded on tape. This really was conversation with alternate answer and reply, interrupted from time to time with personal identification signals.

Other tests took this further. The engineers conceived the idea of allowing Dash to listen to a tape recorded earlier by Doris. Dash began answering the calls as usual, but after 17 minutes 53 seconds, he lapsed into complete silence. On the next and following days the same experiments were repeated with the same results: Dash "rang off" each time at exactly the same place on the tape.

He had obviously realised that it was not Doris on the line. But how? Just before Dash stopped speaking to the tape recording of Doris, this tape contained a series of new sounds appearing for the first time. It was at this precise moment that the porpoise's suspicions seemed to have been aroused.

"If this is the case," as Vitus B. Dröscher so rightly emphasises, "one has to admit that one is faced with a phenomenon unique in the animal world, because sounds of this sort can have no meaning unless they are related to those sounds which the other speaker has made immediately beforehand."

It is as though Dash, the porpoise, stopped his conversation because he could no longer recognise this abnormal mumbling as Doris's voice.

After such staggering results, we may well hope to be able one day to decipher these submarine transmissions. Unfortunately, it does not seem likely in the immediate future. Seven years have gone by and, to the best of our knowledge, electronics have not yet accomplished this miracle.

Why is it so difficult to interpret the "language" of the most intelligent of dolphins? We must refer back to the work accomplished by Dr John Lilly in 1960. At that time, the research workers at the Saint Thomas laboratory made a startling discovery.

While playing back tape recordings of sounds made by the dolphins, it was noticed that, here and there, words and phrases curiously reminiscent of human language could be distinguished. The explanation was simple: like talking birds (a subject to which we shall refer in a later chapter), dolphins must also have a gift for mimicry. But the dolphins' imitations were extremely rapid and confused. The team therefore played the tapes back again at greatly reduced speeds. After many trials and much discussion, the team discerned English words and fragments of phrases which they had spoken and which must have caught the dolphins' attention.

The animals' pronunciation of these words was so quick that, in order to make them more or less recognisable, it was necessary to play back the tapes sixteen times more slowly than the recording speed! Under such conditions, the difficulties of deciphering the whistles – the true dolphin language – are apparent.

Two contrasting solutions to the problem have been put forward. Either man must teach dolphins his language; or man himself must learn dolphinese. It would appear that American zoologists have definitely decided against the former method favoured by Dr Lilly.

The second method is much simpler. In the opinion of Dr R. P. Busnel and K. Norris, inter-communication between man and marine mammals should be based on a system of whistling. This is not a mere Utopian dream: in the Canary Isles and Mexico the natives have long been known to communicate by whistling. Whistling languages are also said to be used in the mountainous regions of northern India and eastern Turkey. In France, before the installation of the telephone, Pyrenean shepherds used a whistle language which could be heard over a distance of five miles. In this way, they could carry on a conversation from one valley to another.

Given time, patience and competent training, perhaps one day we will be able to establish in this way a bridge between cetaceans and man. Even if we are unable to understand a method of communication whose basis is still a mystery, we might be able to form a closer relationship with dolphins. They seem to hanker after this, but it is to be feared that they are more likely to suffer than to benefit from it.

## If sea monsters could speak

Dolphins and porpoises are not the only sea animals to display an instinctive trust in man. We are struck by the extraordinary ease with which other marine mammals accept our presence, and by their rate of adaptability – greater than any domesticated land animal – to what we try to teach them. Their exceptionally rapid speech is admittedly a serious obstacle to eventual communication. But, as Dr Lilly has shown, marine mammals, by pure chance, have recognised a phonetic similarity between their "language" and our own and have shown willingness to be friendly.

Dr Edward Griffin, Director of the Aquarium in Seattle, has managed to make friends with the most dangerous of all whales, the terrible killer whale. It is to the amiable dolphin as the tiger is to an affectionate dog. Such is the voracity of the killer whale that several dolphins and seals are often found in its stomach. Indeed, the killer whale kills for pleasure.

In June 1965, some salmon fishermen happened to capture a male specimen of this dangerous toothed whale. It was bought straight away by Dr Griffin, who had long dreamed of acquiring such a specimen for study. But there was the problem of transporting this cumbersome creature, some 9 yards long and weighing more than 5 tons, the 450 miles to Seattle. A floating tank which could be towed was therefore built for it, and the journey began. For several miles forty killer whales escorted it, summoned by their fellow-creature's distress signals. Once the convoy had reached its destination, the whale was installed in a creek some 30 yards long, forming part of a natural bay, and was fed over 20 lbs of salmon per day. The whale, which was to become the famous "Namu", began to thrive. Every time it approached the quay for food Dr Griffin spoke to it, imitating as closely as possible the whistling of the captive animal. Having no idea of what he might be expressing Dr Griffin soon noticed that one – and only one – of his whistling noises summoned the monster, whose dorsal fin would then slice through the waves with the grace of a sailing ship.

Convinced that he had chanced upon a means of communication, Griffin applied himself tirelessly to imitating the whale's noises. One day he plucked up courage to enter the water. The following day, he continued his experiments from a rubber dinghy. Less than a week later, the courageous zoologist was joining in the frolics of this "assassin of the seas". To sum up, a month after its capture, Griffin succeeded in "riding" astride the great Namu. Wedged between the two pectoral fins of this unique living outboard vessel, he sailed along in complete confidence!

As for verbal communication, Griffin and Namu are still at an onomatopoeic stage, consisting of murmurings and simple whistles, meaning: "I'm hungry", "No, that's enough", or "Come here". However, the American zoologist is certain that he and his killer whale will understand each other perfectly in the end.

In the meantime, one cannot help being puzzled by one detail of this amazing adventure. Thirty to forty killer whales had flocked together in response to the distress signals of one of their fellows. They had swum around the two boats for quite a while, giving every sign of shared distress. And this was because one of these mighty cetaceans, caught in a net and struggling for several hours, had alerted them by an underwater sound signal which we know can be picked up over a distance of several miles. As long as his comrades were near him, Namu, enclosed in his floating tank, never stopped thrashing around and giving a continuous shrill whistle of distress. It was only when he saw the last of his comrades plunge into the depths and abandon the chase that he suddenly fell silent.

A similar phenomenon of non-uniform behaviour had already been observed by Aristotle among a school of dolphins in the region of Tarento. For a long time the authenticity of such tales has been doubted, but truth is often stranger than fiction.

# A dolphin S.O.S.

One of the best attested examples of dolphins' social behaviour was that witnessed on Friday, 26 September 1969, at 6 a.m., by the crew and guests aboard the *Coriandre* (belonging to M. Réaubourg). She had left Port Canto at Cannes at the same time as the *Silène*, on an organised tunny fishing expedition.

I saw the film made on that occasion showing every detail of the scene, but I have chosen to quote at length from the statement made by the principal witness, Patrice Alfieri:

"... Weather marvellous, calm sea. The course is 120°; we are in constant contact with the *Silène* which is accompanying us. About ten o'clock the crackling voice of the walkie-talkie is heard: '*Silène* to *Coriandre*. A large number of dolphins persist in following me ... right in my wake. STOP. They are making fishing impossible. I am changing course to 40° to get away from them. ...' M. Réaubourg replies: 'Received loud and clear. I am continuing on course 120°. STOP. We will meet up with you at noon. ...'

"... The fishing was not bad and after six hours at sea the passengers were anxious to take a rest. Towards noon, M. Réaubourg noticed a great disturbance on the surface of the sea about a mile from the *Coriandre*. Soon it was observed that a baby dolphin, about $2\frac{1}{2}$ to 3 feet in length, was being supported on either side by two adults who seemed to be teaching it to swim. The group advanced to touch the hull of the ship. We had stopped 50 yards from the scene, marvelling at this strange sight and at the same time filming it.

"But events soon speeded up. Some thirty dolphins appeared and making straight for the *Coriandre,* began to cavort about, turning somersaults and leaping from the water, etc. It was then we understood that the little dolphin was in difficulties and that the whole school was acting like this to attract our attention. . . . Suddenly the baby moved off from the adults and started to swim on its own. Very quickly it showed signs of fatigue, swimming desperately with its head above the water. We leaned over the rails calling to it as one might to a dog: 'Come along, little fellow, come here . . . quickly!' all the time encouraging it with gestures. . . .

"Then the unbelievable happened: guided by our voices, the little dolphin changed course by 90° and, making pathetic efforts, hurried towards the boat. The sight of its agony became unbearable. It turned on its back and remained motionless, except for convulsive shudders. It was not more than 10 yards from the boat and was clearly weakening . . . I quickly put on my mask and flippers and dropped into the water to help it. All the adult dolphins now circled round, but somewhat more calmly. I swam gently up to the little creature, quickly lifted its head and held it above the water. Five fathoms below me an enormous dolphin (probably the mother) swam continually to and fro. Unfortunately, just as I was about to come aboard the *Coriandre,* I felt the baby's body finally shudder and go still. I could not help shouting to the boat, 'It's dead!' All the passengers noticed that at the exact moment of the baby dolphin's death, all the other dolphins dived simultaneously and disappeared from sight. . . . Coincidence? I don't think so, since they had been gambling round the boat for more than an hour and a half. Once the little dolphin had been brought aboard it was easy to see that a reddish tumour (in all probability congenital) was obstructing its blow-hole and preventing it from breathing. We were very moved, and concentrated our attention for quite a time on the little body, each one in turn trying mass-age and the kiss of life, but all in vain."

And even more incredible: "On the following day, at the same time, the *Coriandre* passed over the same spot. In ranks of six, almost fifty dolphins escorted us for more than half a mile. With their heads above the water, they went on behaving in an excited fashion which we could not interpret. When M. Réaubourg appeared on deck and leaned over the rail towards them, their behaviour became positively delirious. . . . 'Surely,' said the captain, 'never before in living memory has that been seen.'"

It is incredible that, in this remote watery world, examples of such highly evolved intelligent mammals have been living for thousands of years. Belittled or ignored by men all this time, they are suddenly pre-senting us with innumerable questions.

Ancestors of cetaceans existed on land long before they took to the depths of the ocean. There are many scientists who claim as proof of this the vestigial tail and the hairs on the rostrum ("beak") of the dolphin embryo. Did the cetaceans know the first men so that they recognise us instinctively and is this why certain marine mammals (in this case, the friendly dolphin) seem to understand us better (and certainly more quickly) than so many of the higher animals? But much of the biology of dolphins still remains a mystery. Their sense of taste is beyond question, but they have hardly any sense of smell. Some observers consider their eyesight amazingly good, others hardly worth mention. Their hearing is so acute that they can hear frequencies of between 400 and 200,000 hertz. But none of this explains the extraordinary accuracy of this "natural" which can discriminate so finely and so rapidly that (in comparison) the famous sonar system of bats seems the most ordinary of transmitters and ultrasonic detectors.

It therefore follows that the indecipherable "language" of dolphins might equally well be on a similar scale to that of their detector system, which is so complex that even the most modern electronic specialists have failed to discover its secret.

## Notes

[1] Lilly, J. C.        *Man and Dolphin*, Gollancz (London, 1962)
[2] Lilly, J. C.        op. cit.
[3] Dröscher, V. B.     *Le langage secret des animaux*, R. Laffont (1969)
[4] Evans, W. and       "Observations of the sound production
    Prescott, J. H.     capabilities of the bottlenose porpoise", *Zoological New York*, XLVII (1962).
[5] Dröscher, V. B.     op. cit.

# CHAPTER 5

# The disconcerting amphibians

The appearance of amphibians on the scene goes back hundreds of millions of years. At that time, there were still no snakes, and birds and mammals did not appear until much later. Since then, amphibians, able to live either in water or on land (depending on their stage of development), still have not decided which road to take. Breathing first through gills like fish, they then undergo a profound transformation which enables them to live on dry land. For this, they use lungs.

Of the three orders of amphibians, we will leave aside the Apoda (i.e. the silent, limbless slow worm) and the Caudata (a genus which includes the salamander) and confine ourselves to the Ecaudata to which all the frogs belong. Today 3,000 species of treefrogs and toads are known, living under every sky and in every climate, except the polar regions. Are they as unintelligent as they are said to be, and as poisonous? By and large, all such suppositions are false. The most poisonous toad is soon tamed and develops such a sense of trust that it will never use against its master the natural secretions provided by nature for self-defence.

It is therefore difficult to understand why so many eminent naturalists have shown such a marked distaste for them: Gessner, Cuvier, Bony de Saint-Vincent, La Cepède, Claude Bernard, for example, quite unjustly hated the toad family. It needed all the personality of a Jean Rostand to "debunk" the myth of the "spotted, repulsive and poisonous toad", and to interest the most delicate minds in the mysterious habits of the frog.

Today, hundreds of scientific notes and observations are written in laboratories. The daily development of frog ova has been recorded, their larvae have been vaccinated; they have been subjected to radiation and operations; attempts to discover the secrets of sex-changes in adults are being made. But more and more importance is also being attached to the significance of their information signals.

Every kind of "communication" can ultimately be defined as utterance and reception. When batrachian signals were first classified, sight and hearing were quickly found to be their most important senses. A striking example is the visual display of the European fire toad.

The presence of this very ordinary dark grey frog floating among the aquatic plants of a pond may not be noticed, but once it is emotionally aroused, this discreet creature undergoes an immediate change. Arching its back and throwing up its head, our fire toad stiffens its limbs, displaying the orange colour of its chest and belly and at the same time revealing the equally bright underside of its four feet. Its motionless and fearsome appearance signifies "Beware! ... danger!" – a silent warning sufficient to keep all predators from risking contact with the irritating and poisonous secretions of our big fat gentle friend of a moment ago.

Although its actual visual powers may be only mediocre, the toad, which drinks through its skin, also sees with it to a certain extent. Jean Rostand emphasises the fact that the toad has paroptic vision.[1] Even when blinded, it will make for a light which is not too intense. If placed between two sources of light, it will behave differently, according to whether it can see normally or can distinguish nothing. With normal vision it will make towards one or the other source of light, but if deprived of its two lenses, it will move forward taking an exact middle course between them.

It is the audible signals in particular which interest modern science. Batrachians are certainly not charming singers. Yet one might be surprised if one listened attentively to the various tones and modulations of frog-song, so cursed by poor sleepers on long summer nights.

The sounds made by males resemble bleats, clicks, tolling bells and hammers striking gongs. The female's piping and twittering sounds are so soft as to be almost inaudible to our ears. Often outside our wave-range, they can be demonstrated only by the use of a sensitive oscillograph.[2]

The colours of batrachian "costumes" are as varied as their voices. For instance, in the Philippines, the frog, familiarly known as the "frog with the voice of a cricket", owes its name to the unmistakable resemblance of its song to the metallic noise made by this insect. The Colombian horned frog, when frightened, makes a loud panting noise –AH–AH–AH– which it

59

keeps up as long as the danger lasts. The Cuban tree-frog snores like a grampus; the male arum frog from Africa gives a piercing TCHI–TCHI–TCHI call.

All these contrasting noises have been recorded on tape, either collectively or individually, and subsequently played back in order to find corresponding responses. It is not difficult to imagine the mass of data required to decipher these signals accurately.

Thanks to Madeleine Paillette's original work[3] in this field, it has been established (either by filtering or editing normal sounds or by creating entirely synthetic ones) that phonetic responses are evoked according to the length or pattern of repeated sounds and changes of volume. These responses confirm that the male common French green frog has a range of four distinct types of sound signals. In their order of frequency, they denote: summons, challenge, physical contact, or intermediary signals.

In America, Professor Capranica of Cornell University has done comparable research using electronic recording equipment. He has found a range of six different sounds, each with an exact meaning: the love song, croaking to denote occupied territory, distress, warning and identification calls between two males as well as the cry of pain described by Claude Petter. This open-mouthed cry is unique in its sudden sharp intensity. It is so unexpected that the predator more often than not promptly drops its prey in surprise. For this reason, the cry might well be regarded as an alarm signal to other batrachians in the neighbourhood.

Similarly, in France, F. Lataste believed he had already found in the "song" of toads two series of three different notes according to the age and size of the frog. The first series consisted of the notes

E, D and C; the second, F sharp, G and A. But these musical observations were not decoded successfully.

Professor Capranica, on the other hand, maintains that his own deciphering convinces him that frogs of the same species, but living in different regions, do not speak the same dialect. Male frogs from Texas are not interested in female frogs from Georgia, and vice versa. Taking Madeleine Paillette's research further, the professor found that certain signals were more stereotyped than others. While challenge or contact signals are given in response to visual or aural stimuli, the calling signal may be emitted without any stimulation from other frogs. Depending on the species, the mere sound of an engine, the ticking of a metronome or the whistle of a distant train are in fact sufficient to alert male frogs or deliberately mislead them.

Doris M. Cochran,[4] for her part, has shown that female frogs are insensitive to sounds of less than 50 hertz or more than 10,000 hertz. (The human ear can pick up sounds from 15 to 15,000 hertz.)

Claude Petter, who has made a French adaptation of her work, describes how a tame green frog would begin to sing as soon as a blaring fire-engine passed in the street. The frog expanded its throat to half the size of an egg and answered this false alarm as long as it was audible.

What strange troubadours, so like caricatures of deep-sea divers or of pop-eyed spotty monsters! If not deaf, batrachians are singularly blind and selective where their love-song is concerned. From 15 March to mid-April (or mid-June, according to the species) the frog's flute-like trill can suddenly be heard as dusk falls. First one replies, then a second, then a third joins in, until from all sides comes the

61

song of these solitary ventriloquists! As far as we know, this song, which seems to us so monotonous and difficult to place, is amazingly clear to the female. She listens for a few minutes and then quickly locates the source and position of the summons.

On the subject of communication between males, we should not overlook the fact that batrachians are extremely inept in their mating. The process begins when the males touches the female with his thorax and fingers – and such sexually sensitive fingers too! But sometimes space is lacking, and then several males, holding on to the same female, can mistake the object of their desire. According to Capranica, a special croaking noise is enough to prevent this kind of error: "Take care, chum, you're making a mistake!" The frog comes to his senses and immediately makes off.

What curious amphibians these are, without memory and passion, and what curious males, whose sexual ardour is so oddly threatened by a latent bisexuality. These strange procreators, to whom the female will entrust all the care and surveillance of her eggs, are a curious shadowy link between fish and monkey. Despite their four limbs, their fingers and vocal chords, these amphibians express themselves badly in both gesture and song. Their globular eyes make them seem both startled at their own daring in having one day left the watery world to jump about on dry land, and full of regret for this move.

# Notes

[1] Rostand, J.     *Crapauds et libellules*, Stock (1970)

[2] Rostand, J.     op. cit.

[3] Paillette, M.     "Conditions bio-physiques du déclenchement du signal sonore chez *Hyla meridionalis*", *S.N.P.N.A.F.* (Paris, 1970)

[4] Cochran, D. M.     *Living Amphibians of the World*, Hamish Hamilton (London, 1961)

# CHAPTER 6

# Twenty thousand singing and calling birds

Fifty million years ago by some accounts, a hundred million years by others, an odd creature, the ancestor of all birds and known to us only by a few fossils, appeared on land and in the skies. This was the Archaeopteryx, which was about the size of a chicken, had teeth, short arms with three clawed "fingers" and a long reptilian tail, and it glided rather than flew. We do not know if it was mute, but very few birds we know are mute.

Apart from the New Zealand kiwi, dominated by the sense of smell, the silent American vultures and the flapping stork, birds of every feather and from all countries call, whistle, scream, cackle and sing in a hundred different ways. They express joy or fear, loneliness or sexual desire. They warn their companions of danger or call to them for help; they jealously proclaim territorial rights; they attempt to intimidate or seduce – and so on. Some are even privileged enough to "speak" the language of man but unfortunately without meaning.

We must add to these vocal displays the language

of the body, whose postures and gestures are means of communication among our feathered friends as among all other creatures throughout the world.

These various signals – calls and other vocal sounds, looks, attitudes and beatings of wings – are inherent in the new-born chick. Innate and hereditary, they are the natural means by which each newcomer communicates with its family. Later, they are means of expressing sexual activity. But are these gestures and vocal sounds stereotyped so that, for example, a crow from the French plains could make itself understood by a crow from the tundra of Arctic Russia? We have no idea.

To survive, the "feathered family" have sensory means at their disposal common to all vertebrates, plus a few peculiarities all their own. Birds' sense of smell is not highly developed. Their hearing is little better than our own, while their ability to taste is practically nil.

On the other hand, their sight is exceptionally good, and each eye has one or two central foveae. There is a slight difference between an owl's eyes and a chicken's (depending on whether the eyes face forward or to each side of the head), but the slightest movement of the head gives all birds a panoramic field of vision with perfect perception of form and colour.

**Aural or visual creatures?**

The hearing of song-birds is extremely selective, but we can point out at once the importance of hearing in the most innate and dominant means of communi-

66

cation in birds: those between a mother and her chicks.

A hen, for example, who will be quite distraught when she hears the chirping of her chicks, will show no anxiety at their plight if they are placed under a glass cloche simply because she cannot hear them. W. and M. Schleidt of the Max Planck Institute have proved by experiment that this inhibition of the maternal instinct depends on the suppression of hearing and sound signals. When turkeys had been rendered deaf by surgery to the inner ear, it was observed that they incubated their eggs normally, but no sooner were they hatched, than they killed one newborn chick after another. Then one of the deaf turkeys was given artificial eggs to incubate and then shown a newly hatched chick. She looked at it without the slightest sign of maternal concern and, as the chick approached, she ruffled her feathers, hissing angrily, and tried to peck it ruthlessly as soon as it was within reach.

Birds, therefore, seem to be quite unable to visually recognise their young. The protective instinct prompts mother hens and turkeys to attack anything which comes near their nest. They do not attack their own offspring with the same violence because the chicks' cheeping, as long as it can be heard, restrains or cancels out their aggression.

Apart from this particular case, linked with the maternal instinct, it seems to be the sharpness of their vision and the speed of visual stimuli which play the most important part in everyday communication between birds.

With the exception of song-birds, the "vocal language" of birds seems to be less important than the "language of gesture". Watch a hundred starlings

67

in flight, and note how the whole flock takes flight, changes course and decides to land at the same split second. Each starling's eye is more sensitive than the most sensitive of rapid camera lenses.

Even more curious, when two pigeons arrive together on the same ledge, they will settle a few inches apart. If one of them decides to take off again, the other stays put, a lightning glance having informed it that its companion is not going far; it will even know exactly where the other will land.

This power of visual discrimination has inspired laboratory experiments, and we now have proof of birds' ability to differentiate between colours, and to pick out objects or drawings from others of similar size and shape. However, there is a lot of work still to be done on birds in their wild state. In this respect, modern techniques have already given valuable results. Telephoto cine cameras have demonstrated that the great gliding birds (such as the eagle, the gull or albatross) only need to make an almost imperceptible movement of their wing-feathers to make maximum use of air currents. Observations of behaviour and subsequent experiments are, nevertheless, far from easy with birds which manoeuvre so quickly and at such heights.

Now that we can make sound recordings, we are better informed of the oral communication between members of the same species. By playing back such recordings to various species, we can identify them, since both recordings and natural vocal sounds produce the same reactions. A hen makes a wide variety of sounds. When laying eggs she clucks rapidly, ending on a brief but more strident note. There is the quiet call to assemble (the family call), the more dramatic alarm, defence or attack signals, and so on.

But different species respond in different ways even to the alarm signals, according to their tone.

As soon as a crow has given this signal, the whole flock flies off cawing, but in the case of tits all will immediately "freeze" and fall silent; even the young will obey and stop their twittering. This is not, then, a simple call. According to its pitch, it can have much greater meaning indicating whether the enemy is near or close, in the sky or on land. Grimsell has even noted that, at certain times, birds may react to this signal in different ways. Should a hawk approach a completely silent colony of tits, all the birds, young and adults alike, will immediately set up an infernal twittering, lasting barely two minutes, and will then relapse into silence. According to Grimsell, the explanation is not fear, but the equivalent of "scrambling" on radio or telephone, thus making it impossible for the attacker to locate its prey.

At one time a distinction was made between human language and the sound communications between animals by stressing that the former was "learnt" and therefore variable, whereas the exclamations, calls, whistles and other animal sounds were stereotyped and unchanging. Today we know that this is by no means the case.

After recording the distress signals emitted by crows, Dr Burnell and animal psychologists at the Institut National de la Recherche Agronomique in France tried an interesting experiment. In the immediate vicinity of a French rookery they relayed recordings of the distress calls made by American crows. There was no reaction. One must therefore assume that in different countries crows "speak" their native dialect, which differs from that of other similar species of crows, brought up in other climes.

However, what applies to crows does not apply to magpies, although they are related. R. de Joly,[1] during his visit to Mexico, spotted two magpies in a tree, and gave the "summons" call of French magpies. In a few moments magpies arrived from all quarters. They settled on the palm trees lining the road and were joined by more and more others, all in a state of extreme excitement. With wings outspread and half closing their nictitating membrane (the inner protective eyelid peculiar to certain animals and birds), they swooped towards him. Somewhat alarmed by the noise and this ever-increasing gathering, de Joly hastily got back into his car.

When he repeated the experiment at the same time next day, the results were just the same. On the morning of the third day he returned, but this time without making any call. Lots of magpies were there to meet him; they had recognised him and now escorted him in the same state of panic as on the previous days. Making the summoning call proper to this species had worked, yet this was a different race from that in distant France.

Like crows, seagulls do not speak the same language in different coastal regions. This was easy to observe when their cries of alarm were recorded and then relayed by loudspeakers near airfield runways where the birds were causing an obstruction. Recordings successful on one continent were ineffective in another.

## Safety for the smallest

Every day fresh progress is made in the study of bird sounds. Thus A. Brosset and C. Chapuis of the Laboratoire d'Ecologie in France have discovered delicate adaptations in the sounds of very young birds which have left the nest at the age of ten to fourteen days. Scarcely able to fly, they remain on the ground at the risk of becoming the victims of animal predators. Nature, which has adapted them to their circumstances (their dull plumage blends with their surroundings and at the slightest hint of danger they remain completely motionless), goes even further. During this period the fledgling depends entirely on its parents for food. Its cry for help must, therefore, have maximum range and penetration for them to hear. But such a cry could well betray its presence to the many predators. Thanks to natural selection, the cry for help made by the hungry little bird is such that even the most highly trained ornithologists can pass close by without hearing it.

Furthermore, on the rare occasions when this call reaches our ears, it is almost impossible to locate it. It has been suggested that the human ear is attuned only to certain wave-lengths not within the scope of the cry emitted by young birds, and it therefore seems possible that the fledglings' natural enemies are also unable to pick up their chirruping.

Consequently, tests were conducted with blackbirds, larks, buntings and sparrows (both young and adult) and with predators such as dogs, foxes and tayras (a South American member of the genus Mustelidae, resembling a wolverine or marten). The reactions of these carnivores to the various taped calls of certain species of birds relayed at the same level of amplification were then compared.[2] The dog, the tayra and the fox reacted perfectly to the alarm call of the adult birds, but the S.O.S. of the youngsters, relayed, remember, at the same level of amplification, appeared not to affect them. This is thought to involve not a faulty hearing of the call, but a filtering of the stimuli applicable to particular sounds. Research is in fact continuing. Only by unbiased observation of animal behaviour and by controlled experiments can we hope to find a solution to such problems. With luck, such research occasionally leads to the most unexpected results.

## A scientist who speaks "bird language"

Only because Professor Konrad Lorenz had the infinite patience to watch the hatching of a grey goose's eggs and then warmed and fed, with his own hands, a newly hatched gosling did the idea of the impregnation principle occur to him.

A long time could be spent discussing what separates the concept of "communication" from that of "true language". But was it not by means of her language, vocabulary and mimicry that Lorenz's little gosling instinctively revealed to human intelligence one of the most unassailable of biological laws?

Today this story is widely known: how, after her very first contact with what was to her the outside world, the gosling felt that Lorenz was abandoning her when he tried to slip her under her mother's belly. She absolutely refused and tried clumsily to escape, with such cries of distress, that the tender-hearted professor took her once again in his hands and carried her to his room.

From that moment the die was cast. Lorenz adopted his gosling, calling her "Martina", and gave her protection, lodging and food under his own roof. Did he for a moment suspect the consequences of his action, or foresee the important part that this fragile little creature was to play in his life?

During the first night the link between Martina and Lorenz was forged when the professor was suddenly awakened by an urgent whistling noise. Hardly had he leaned towards the cushion where he had settled Martina than he was greeted with gentle murmurs full of affection and joyous recognition. As soon as he moved away to return to bed, Martina, thinking herself abandoned once again resumed a plaintive vi-vi-vi-vi ... vi-vi-vi-vi ... vi-vi-vi-vi ... (Where are you, where are you?) This sound was repeated until Lorenz, who, thanks to recordings, knew something of goose-language, gave a reply which was unconsciously onomatopoeic: "Gan-gan-gan!" (pronounced as in the French fashion) which signifies more or less, in grey goose vocabulary, "I'm here, I'm here". Miraculously Martina was reassured, calmed down and silenced. For the first time man had made himself understood by an animal by speaking its own language. Lorenz had, in spite of himself, become "Mother Goose"!

From these beginnings, he applied his whole heart and mind to the work, so that today Professor Lorenz is as well known as Professor von Frisch with his bees or Dr Lilly with his dolphins.

But Lorenz did not confine himself to a study of the vocabulary of geese. He also spoke "crow-language" or, more accurately, the idiom of jackdaws. His successive discoveries and failures with jackdaws were enthusiastically followed by the press. One of

73

his tame subjects showed real affection for him and insisted on trying to feed him as it would have fed one of its own offspring or a mate. In such cases, members of the crow family use their tongues to thrust larvae of weevils and insects, impregnated with saliva, deep down the throat of their young. Lorenz's tame jackdaw tried to fatten him up in the same way. Meeting with a firmly shut mouth, it then sought other means of entry and attempted to push the food into his ears and nostrils!

Professor Lorenz has not yet determined whether "jackdaw language" consists of more "kias" sounds than "kiaws" "two words" which in crow-language seem to mean "come with us"). What is certain, however, is that this painstaking observer has been able to distinguish clearly between calls of hunger, pleasure, fear (as yet not clearly defined) and panic. These are instinctive sounds which come naturally to fledglings and which they use later, just as their descendants will know and use them.

As we shall see, the musical song of birds is less rigid and acquired in a much less precise fashion. The basic principles are innate among nearly all the sparrow family, but their song develops and is enriched by learning, just as the babbling of the child eventually becomes adult language.

**Birds proclaim their love**

Facial mime in most birds is negligible with the exception of the crested variety. Wattled, tufted and plumed birds raise their feathers when frightened, angry or threatened, or, like the turkey, infuse their wattles with blood.

The general sexual display (whether or not accompanied by sound) is much more familiar and obvious to us. Birds, especially the males, undergo a real transformation at mating time. The non-singing birds make up for their limited vocabulary by the most spectacular ceremonial gestures. Dancing, the universal code which expresses emotion, comes to the aid of the no less universal code of music.

To the accompaniment of utterances which may seem to us unpleasant, non-singing birds indulge in contortions and postures as varied as they are bizarre. In fact, this behaviour is activated by hormones, and is necessary to arouse the female to a state of receptivity. The male's general appearance is so extraordinarily different that it might be difficult to believe that the male's behaviour is not a conscious effort or a "wish to please".

Experiments conducted by Matthews have thrown further light on this impressive phenomenon. Not only does the strutting of a male pigeon before a female excite her sexually, culminating in ovulation, but this visual stimulus is entirely impersonal. A hen pigeon's own reflection in a mirror will have exactly the same effect on her. The words auto-eroticism and narcissism immediately spring to mind here, and indeed, everything points to this. One could well believe that a peacock is aware of the splendid eye-like markings on its spreading tail; that a grouse takes pleasure in its rich plumage and that the syncopated rattling, crackling noise he makes while strutting about excites him as much as the female.

Each species has its own repertory. Like a great ninny, a farmyard goose will simulate a furious attack on the first owl that comes within its reach and will then loudly trumpet its hollow victory. Ducks quack and, like old seasalts, roll from port to starboard; crows caw (or croak) and fluff out their plumage; waders (cranes and herons) swell their necks and gabble; the eagle screams and suddenly spreads its wings; cocks rise on their spurs, spread their ruffs and crow as loudly as possible; the gentle turtle-dove moans and the pigeon coos; the puffed-up turkey gobbles; the peacock shrieks; the quail chatters; the jay wheedles; the partridge croaks; and the plover cries like a child and with lowered head performs a thousand dizzy loops in the sky. Finally, if some owls screech and others halloo, all nocturnal birds of prey can hoot.

Among certain species these demonstrations sometimes take on a symbolic meaning. The bower bird decorates its nest with coloured pebbles, pieces of glass, petals and flowers. Others offer gifts: a hawk will bring a mouse; a tern, a small, freshly caught fish. Some will first busy themselves in the strangest manner to establish their territory. The grey-headed, great spotted and green woodpeckers strike a dead branch with their beaks as though beating a drum, each species employing its own particular rhythm and frequency.

The mating display of swans and emperor penguins involves both sound and gesture and follows a completely different pattern.


**The not so gentle swan**

With the coming of spring swans have already chosen their mates after merciless battles with rivals. (Among most other birds such fights are mere pretence.) After beating each other with their wings, sounding like cracking whips, and making hammer-blows on the head with their beaks, neither combatant will yield until he has encircled his adversary's neck with his own. The victor then deliberately forces the head of the vanquished under water, maintaining it there until his unfortunate rival, suffocated or drowned, lets go.

These gestures of violence are now transmuted into shows of tenderness towards the mate. The amorous display of swans starts with a sort of ballet when they glide towards each other, heads held high, then suddenly draw apart with snuffling noises and tender murmurs, and then all at once, they entwine their necks as if to fight. What was a moment ago, between males, the embrace of death is now the symbolic tying of the marriage knot! The finale is a silent duet, their softly lifted wings resembling the billowing sails of a white yacht gliding towards Paradise.

## The serio-comic penguin

Dressed in black with white shirt fronts, emperor penguins, like swans, present a fine appearance. But can we take these bizarre birds seriously when in response to their sexual instincts (that is to say, to mate, reproduce, hatch out and bring up their young), they choose the most frozen, solitary and inhospitable place on earth?

To see them gravely waddling along, day after day, with nothing to eat, in order to instal themselves on the most uncomfortable of ice-floes, huddling together and chanting monotonously, it is inconceivable that they can then indulge in tenderness.

Nevertheless, as Drs Sapin-Jalouste, J. Prévost, E. Pryor and other Antarctic biologists have observed, emperor penguins show the same sort of agitation and eloquence as birds inhabiting the balmiest island.

They never display aggression. However faithful they may be by nature, they never fight over a female. If a couple meet, they greet and introduce themselves. It is all very correct and not at all flirtatious. The male raises his head and then bows. With neck and beak pointing to the ground he murmurs something in a low tone of voice. The female listens politely to the end of the discourse, then discreetly takes herself off while the male triumphantly slips away to join the rest of the band in the plainsong of the assembled males.

From time to time this continuous chanting is interrupted by a loud, sharp cry, like the identification signal in an S.O.S. Since it is almost impossible to distinguish each other in the Antarctic night and in howling blizzards, it is by this cry that "married" couples call to each other and reassure each other of their presence.

But how different is their mating act compared with what we know of other bird life! Here is what Dr Prévost, a specialist on the polar regions, has to say of a couple of penguins, paired the previous year, who now suddenly meet again through this unusual oral identification: "With the neck a little dilated they stand still in front of each other, shifting their weight from foot to foot, the body bent. The head is slowly lifted to the sky, the base of the neck dilates more and more, causing a slight fluffing up of the feathers. . . . The two birds may then either stay pressed together, breast to breast, or remain slightly apart. Their immobility is almost complete; their eyelids, half closed, may then blink a little; their visual and aural awareness of others around them seems to be greatly reduced. . . ."[3]

One is tempted to say, ironically, of this enraptured couple that, even in the polar regions, lovers become lost to the world.


## Birds are not robots

Seeing these penguins reacting to an apparently fixed instinct, we might well think that birds are devoid of intelligence.

Professor Rémy Chauvin has very clear-cut views on this point. "Until quite recently", he writes, "it was believed that, because the cortex of birds is less highly developed than that of mammals, their behaviour was entirely instinctive. Today, opinion is quite the reverse. It is generally admitted that as far as the complexity of their behaviour is concerned, birds lag only a little behind the higher mammals."[4]

There are many examples to illustrate this authoritative opinion. On very rare occasions, during the course of "love duels", which seem to be dictated solely by sensory reactions, real flashes of intelligence have been observed, which seem to indicate genuine feelings. There immediately comes to mind the story of the two storks, which has since become a disquieting symbol of conjugal fidelity.

In the early days of spring, the female, wounded by a gun shot, managed to return to her nest and lay down. During the time needed for the broken bones of her wing to set she stayed on her eggs and when they had hatched, still unable to move, she watched the first attempts of her young to fly. Meanwhile, the male provided food for his wife and family. When the time came to leave on their yearly exodus, he took off, followed closely by the young ones, but without the mother. He had encouraged her in vain to rise and join him and the other migrating couples. When she saw him leaving, she rose up on her feet, vainly beating her wings, all ready to join the rest, but was forced to give up. It was now that an extraordinary thing occurred. Already high in the skies, with his legs suspended and ready to join the ever-increasing number of migrants, her mate suddenly decided to resist the ancestral call of his kind. He abruptly changed course, swooped deliberately down to the poor invalid who could only clack furiously to proclaim her distress, and rejoined her near the nest.

Autumn and winter came and went. The strange saga of this devoted couple had touched the whole village. There was a dramatic end to the story. When the fine weather returned, the female attempted the impossible. Leaving the nest, she managed to reach a nearby roof top, then, risking all, launched herself into space only to crash to the ground.

What can be said of the various stimuli which alerted the male for departure, then prompted him to halt in mid-flight and led him to decide – at the last moment – to remain beside his mate? Certainly her distress call must have played a part, but what secret process created such a profound "self awareness" that this male should over-rule the automatic, compulsive instinct which governs all migratory birds in favour of parental feelings towards his companion? We still know very little of the methods of communication that exist between migratory birds.

## In-flight communication

Gregarious birds which fly in their thousands without orders and in no definite formation inevitably bring to mind immense, ungoverned shoals of fish. But organised formations in either sky or sea present strange problems. The classic example of the "V" formation of wild duck is a case in point.

As soon as the leader slows down and descends a few feet, his place is promptly taken by one of two ducks who flank him at a short distance to his rear. However, this is not an absolute rule: any other duck in the formation can take his place. Here, again, one asks: "Who chooses this replacement?" And how is the order given to relieve the leader?

Professor Hubert Frings[5] maintains that continuous consultation occurs during the course of all organised migrations. When cranes, for example, take off, are in full flight or ready to alight, two clear and very different cries can be heard. The first, sharp and abrupt, has a forceful meaning which might be translated as "Let's go!"; the second, a plaintive, shrill note, expresses fatigue or hunger ("Shall we stop or go on?"). But here we may become sceptical, for, according to the same authors (Hubert and Mabel Frings), the decision to land or to continue is taken by a three-quarters majority of the birds. To prove this, the Frings from their position in the middle of a field relayed a recorded "call to rest" to a flight of cranes.

This specific signal (which, as we have seen, crows answer so well) should have brought the whole flight to earth. But not at all! It did not disturb the cranes for a moment. The Frings repeated the experiment a dozen or more times but with no success. According to Vitus B Dröscher,[6] reporting these events, the call to rest succeeded only once. On only one occasion did the shrill voice of the loudspeaker seem to tip the balance. Indeed, almost immediately, the cranes landed in the immediate neighbourhood of the tent which hid the experimenters from view. But, maintains Dröscher, this was no doubt because the necessary majority had been obtained and the decision thus taken for the whole flight.

Such claims naturally leave room for doubt. And yet, it must be acknowledged that in all bird communities we find the same problem: either they are under orders from a leader or else decisions are reached by mass consultation.

Sea birds, in particular, congregate in enormous groups on cliffs or inaccessible islands. Every year, successive generations gather on precisely the same dates. But it sometimes happens that tens of thousands of gulls, assembled quietly on the ground, may be suddenly alerted, without any apparent reason, by the excited behaviour of one of the colony which has thoughtlessly given the alarm call.[7] Then, as one, the whole colony will take wing, but quickly realising that there is no danger, they will circle round a few times and then return to their starting point.

A colony can be deceived in this way two or three times, but rarely more. Furthermore, if their too-excitable fellow gull again gives a false alarm, he will be rendered silent or driven from the colony for breaking the rules!

## Talking through an eggshell

It was always believed that birds relied mainly on their sight. Today we know that they are equally dependent on their hearing. The role of their songs and cries in their communication would have already been proof of this, if recent studies of bird colonies had not revealed more extraordinary facts concerning relations between members of the feathered world. Be they gannets, cormorants, or guillemots, each couple occupies its own territory, with an appropriate distance between each nesting pair. Sometimes, many families settle so close to each other that together they make up virtual societies. Yet, among all this vast anonymous crowd every male knows perfectly well its own place and mate and each knows where and how to find them. Similarly, every mother can pick out her chick wherever it may be.

How do these birds recognise their mates, their young and even their eggs?

In an attempt to find out, Tschang and his fellow-workers observed these birds at length on the Norwegian island of Vaeröy. After ringing or marking female guillemots, they dispersed the whole colony by firing a shot into the air. While the birds were out of the way, the eggs, laid straight on to the rock, were marked and their position changed. Once the alarm was over, hardly had the birds returned or been freed, than each mother-bird recognised her own egg and lost no time in moving it back to its proper place. But more importantly, four days before they hatched, Tschang recorded the cheeping of the chicks inside their eggs and also the voices of their parents. The tapes were then played back to the adult guillemots and to the future generation still

83

inside their shells. Every time, despite the background noise made by a gallery of tens of thousands of birds, the parents and the future chicks reacted only to the call of members of the same family! Thus, in the heart of these countless colonies, broody birds can distinguish the voices of their offspring even before they have seen them and the young know their parents' voices before they have seen daylight![8] To confirm these results, the scientists experimented with loudspeakers on artificially hatched chicks. The experiments were always 100 per cent positive. This is why, Tschang suggests, seabirds will only feed their own chicks. Moreover, the chick itself will only approach its own parents for food, accepting nothing from other birds. For him, neighbours can only be strangers, because they have not the same pitch of voice as his family.

**What do song-birds sing?**

It might be thought that bird-song, like the various gestures described, is also a language of love, except that many of the sparrow family sing the whole year round. Perhaps too prosaically ornithologists see only one essential function of bird-song: a means of asserting territorial rights. In support of this theory there is an excellent study of the robin's song by J. C. Bremond.[9] He discusses how a robin will defend its territory by visual means and sound signals, and he considers appearance, size and method of marking the frontiers of the territory as well as the role of singing-places, the spread of signals over neighbouring trees, and the basis of the song.

Nevertheless, we cannot altogether dismiss the

mysterious artistic quality of bird-song, which has inspired musicians throughout the ages, long before modern zoologists showed any interest in the subject. Indeed, Pierre Pellerin has devoted a whole section of his book[10] to great composers whose enthusiastic interest in the harmonies of bird-song led to direct and unashamed inspiration. "What would the Pastoral Symphony have been," he writes, "without the rustic trio comprising the passionate warm notes of the nightingale, the elusive piercing call of the quail – first introduced by ascending semi-tones – and the call of the cuckoo which fills the void with dreaming melancholy. How often and how attentively must Beethoven have listened to these songs in order to master them!"

Closer to our own day, César Franck, Vincent D'Indy, Gabriel Pierné and Ravel have all fallen under the spell of these heavenly songsters, but it is Olivier Messiaen, dedicated stubbornly to enriching what he calls the "vocabulary of musical language", who has made the most extensive use of bird-song. "Messiaen observed trills and mordents, vocalisations, sostenutos, appoggiaturas and roulades," continues Pellerin. "In the stillness of the dawn he recorded vibrations of ripples and warbles and never ceased to be astonished –although he had known it for a long time – at the incredible versatility of blackbirds, surpassing the human imagination."

Scientists realise the danger in this imagination! Doubtless this is the reason why animal psychologists seem to have avoided pronouncing on the significance of musical trills and vocalisations, on the pretext that we obviously cannot prove whether birds perceive them in the same way as we do.

85

# What is song?

"Song," wrote Professor Grassé[11], "consists of a series of sounds, made up of intervals of various pitches. A bird can perfect or vary the 'phrase' which it has inherited in a basic form, by listening to the song of other members of its species. A bird is influenced in this way from its earliest days as a chick when it is still incapable of singing. It therefore seems to possess a musical sense right from birth, and sings first of all to express its joy."

An instinctive expression of joy? "I am happy"; a vocal equivalent of physical display? "I am the best ... the strongest!"; an expression of pride linked with and confirming a territorial sense? "I am at home here!"; defensive aggression? "Keep away!" – bird-song is all this and more.

In Bremond's opinion, a bird's song conveys its identity ("It's me ... the robin!"), as a species and as an individual. It also expresses its sexual condition, and its degree of aggression, it indicates the presence of an enemy, and defines its territory. All this suggests a form of "language" when the solo is replaced by vocal exchanges between two birds of the same species, and when each waits for the other to finish before replying immediately with a song made up of different modulations and tones. Such is the case of the hen finch, which, according to the poet Lise Lamarre, utters hysterical bursts of song as soon as her male pursues her. Such is also the case of the young blackbird which, every evening among the same tangle of branches which is his domain, flies into a homeric rage with his spouse ("You're late to bed... dawdler ...") because she isn't back home yet. When we see it impatiently hopping about, frantically

86

clacking its beak, writes our sensitive poet, we cannot help making such an analogy.

A similar case is recorded by W. Thorpe[12] concerning communication between pairs of gonoleks in the dense forests of Uganda. Thorpe records that, in order to avoid confusion with other birds in the immediate vicinity, each pair observes between utterances a pause of a certain length which acts as an identification sign. It is extremely difficult for the human ear to register this sort of pause, which is measured in thousandths of a second.

The song-bird, therefore, can signal and reply far more rapidly than a human voice. Experiments made on this aspect by the Max Planck Institute are described by Dröscher.

After placing a metronome close to a shama blackbird (whose whistling song shows an astonishing variety), Johannes Kneutgen observed that the bird was consciously affected by the instrument's rhythm and modified its "tempo" accordingly. When Kneutgen increased the speed, the blackbird adapted its own song straight away. When he increased it still more, the bird tried to follow it, but then gave up the attempt and, instead, sang a completely different melody from its repertoire, and one whose tempo corresponded better to the other rhythm which had so annoyed it.

There are highly interesting experiments still to be conducted in this field. We know that Dr Lilly's experiments to decipher "dolphinese" were hampered mainly by the great speed of these marine mammals' utterances. Perhaps we will be confronted by similar difficulties when we attempt to catch and make sense of the exact components of the trills and warblings of these no less mysterious song-birds.

# The privilege of song

Song-birds are like human beings. Not many of us can claim to sing well – only the privileged few endowed with larynx, vocal chords, a vaulted palate, and a large thorax. Even so, the most natural "golden voice" requires study and training to develop.

Among birds, the natural anatomical structure of their organ of song (the syrinx) with its air sac and vibratory membranes, would not in itself be enough to give finches and nightingales, for example, the extensive repertoire they develop later. In addition to this rudimentary phrase of song which is inherited, certain species also have the gift, peculiar to winged creatures, of vocal imitation.

Ornithologists differ on the nature of this basic, inherited first phrase. According to Oscar Heiroth, if a baby finch is reared in captivity, separated from its own kind, its first notes will bear no resemblance to those made by its brothers in the wilds. Other breeders, in contrast, assure us that everything starts in the egg and that warblers reared in the strictest isolation, will chirrup their family song without ever having heard it. As far as tits are concerned, everyone is agreed. Hardly have they left their parents, at a very early age, than they fly onto the nearest branch and immediately give the alarm cry in the tone of their mother.

However, it is only by imitation or learning that all these halting sounds eventually become songs and, according to the species, very different songs. A warbler, hardly out of the egg and which throughout its short life has heard only the song of a bullfinch, will perhaps sing like a bullfinch later on, but a tone peculiar to warblers will always be apparent in its

88

song. On the other hand, finches, buntings and nightingales cannot really sing until they have developed their inherited powers by hearing their father or any other adult male of the same species.

It is here that confusion is apt to arise.

To be completely accurate, the nightingale does not possess a "unique" song pattern, although it is a song-bird with an individual range, just as among humans there are light tenors and baritones. It has been suggested that the nightingale's song is influenced so greatly by its surroundings and modified by the songs of neighbouring birds that it might even have a dialect. This, however, is open to question. What is certain is that we are not concerned here with a difference of song language, but a difference of tone in the case of individual birds.

C. Chapuis's painstaking research in this field has led to this surprising discovery – that the pitch of song-birds depends on whether they live in northern or southern countries. To prove this, a long and detailed comparative study was made of the song of 20 different species living in Holland, Belgium, France and Spain. Such evidence (made by tape recordings and spectrographic analysis) confirmed what A Brosset, Lorenz and other animal psychologists had already suspected: the further south the lower the pitch of the song; the further north, the sharper it becomes. Does the action of light and warmth affect glandular activity? We do not know, but it is a long established fact that among children's choirs, soprano voices, so common in Europe, are rarely to be found in North Africa.

Another strange example of musical adaptation can be mentioned. A certain species of finch gives squatters (i.e., chicks from eggs deposited in their nest by widowed birds) the same blind welcome as the classic cuckoo receives. But obviously these "squatters" sing differently from their foster parents, and suspicions would be aroused if nestlings, taken to be their own offspring, were heard "speaking a foreign language". So what do the squatters do in such a case? They quite simply adapt – first of all vocally, so that they express themselves in finch-song, and then physically, their plumage taking on, by some inexplicable mimicry, the general colour of their adoptive parents!

We see, therefore, that bird song is not so stereotyped or fixed. Canary breeders hold song schools and song contests. Among canaries selected are malinois, which utter a sharp note; Spanish flautists (which have a high pitch); the German Hazer (a lower tone because they sing with closed beaks), and the smet canary which surpasses them all in expertise. These birds have been taught to cultivate their natural *rourou*, their *blaou-rouli*, then *luluhouhou* (excuse me, this is canary language), and some judges will not accept the *"tilling"* for a *"tilloung"*, or a well balanced *bubu* roulade for a twittering or sound of little bells.[13]

And nothing could be more varied or complex than the *dullu, lullu, lulu, lu-lu-lu-lu, lu-lu, duli, douli, didouli, didouli, dli, dli, duli, dli, dli* ... which the woodlark sings night and day from February to October.[14] With its trills, modulations and different tones, this "nightingale of the moors" has been compared with the most highly gifted small song-birds.

Can such repertoires be regarded as a form of "language"? The answer is "yes" in so far as they form a biological code, a means of communication which helps to protect the individual and the species;

but "no" if we consider it as conscious participation. Bird-song no more greets the sun or the spring than it weeps at the arrival of winter. It is no more a proud cry of victory than it is a cry of defiance or a war cry. It expresses mood, an involuntary psycho-physiological reaction, just as our frowns, laughter, sobs and spontaneous exclamations interpret (often better than words) our states of mind.

All the same, bird-song is music and music is subjectively "language" in that the same notes and harmonies have a similar effect on all those, irrespective of nationality, age, education, or culture, who fall under their spell. At least as far as man is concerned, bird-song, like music, tends to produce the same effects all over the world.

Modern ornithologists are not content to observe bird-song; they have invented a new game which demands patience, an artistic sense and a great deal of love – in a word, they want to record bird-song. As with big game photography the use of acoustic equipment will keep science happy and will be of service to music and poetry at the same time.

Let us hope that such modern use of leisure will encourage modern man to keep in touch with the true standards of life.

## When birds have their say

We must return to Lorenz and what he was able to reveal of "animal language" to understand what separates the goose (which has for some reason become the symbol of stupidity) from the various mimic birds. Where these are concerned, we have no need to talk about intelligence, because their powers of mimicry are enough to guarantee our amused indulgence. Our attitude towards monkeys, whose gestures parody human behaviour, is exactly the same.

From their earliest days, goslings respond to the finest nuances of the adult goose's complex vocabulary and to these alone. Their honking, which is varied and embraces many different tones and syllables, can be interpreted to mean "halt", "move on", "hurry up" and "slow down", or as alarm signals when confronted by serious danger or a less serious threat, etc. In short, unlike other birds, geese never talk unless they have something to say. Konrad Lorenz has good reason to be more proud of mastering the nasal language of geese than of speaking the most difficult human dialect.

Meanwhile, we reach the heart of the problem. Of all the creatures which fly, swim or creep around in this world, only a few birds "speak", yet, paradoxically enough, it is these privileged few which seem furthest from a real language, because they can neigh like horses, bark like dogs or creak like door-hinges and sometimes inexplicably echo the human voice perfectly. Yet these birds have the fewest gestures or vocal signs denoting a specific code. This is one aspect of the mysterious birds, mainly of the parrot family, which are able to reproduce articulate human language. A great deal of attention has been devoted to this subject.

The number of talking birds is not yet definitely established. The last to be discovered is the resplendent lyre bird of Australia. Imitating all the birds of the forest, its principal cry is a peal of laughter. Unfortunately this largest passerine in the world, as P. Henry Plantain calls it, is the most difficult bird to locate. First of all, you think it is here, then it seems to be over there, and even the most learned ornithologists can be baffled as to the exact source of its cry.

But the classic examples are parrots and parakeets. One can well imagine the astonishment of Christopher Columbus's companions when they first saw and heard these "fire birds". A flight of chattering parrots is said to have prevented the famous navigator from turning back when he was on the point of giving up all hope of discovering a new land. But this is merely a legend, since wild parakeets are unable to "talk".

Australian cockatoos, with their little stiffly plumed helmets, content themselves with expressing surprise or anger by spitting. The South American macaw can only utter deafening cries. An exception might be made for the red macaw but its vocabulary is restricted to a few words. Among the most gifted of talking birds are the familiar green "Poll" and the affectionate grey parrot from Gabon.

After the hook-beaked birds, we come to the talkers with straight beaks, whose performance was already appreciated by the Greeks and Romans a century before Christ. Among these are crows, rooks, magpies, jays, canaries, starlings and blackbirds. It is to this latter category that the famous myna bird belongs. This black bird has become fashionable and deserves its popularity for its engaging qualities, its simplicity and the ease with which it can learn new sounds, songs and words, and repeat them accurately. Originating from Java and the Indies, all mynas have dark plumage with a dull green metallic sheen and physically resemble the common blackbird. Their jerky hopping movements are common to all passerines (perching birds). All have the same wicked twinkle in their eye, a long oval head, a pointed beak, and the same curious crest, with its prominent yellow wattles extending over the nape and cheeks.

Myna birds may gradually usurp the place of parrots as household pets. During the Second World War, Walt Disney established their popularity in America by giving a dinner party for one of these birds, called Raffles. It whistled the "Star Spangled Banner" with such patriotic fervour that it sold 15 million dollars worth of National Defence Bonds during a coast-to-coast propaganda tour.

"Talking" birds were known to antiquity long before parrots had ever shown this talent. Pliny tells the story of a famous crow which was adopted by a humble cobbler. Every morning the bird flew to the forum and alighted on the rostra, greeting Tiberius and the tribunes by name as they arrived. Then it returned home until the next morning. But the crow was killed one day by a jealous neighbour, who was lynched for it by the mob. The bird was venerated by all Rome, and honoured with a magnificent funeral. Its bier, adorned with flowers and borne on the shoulders of two black slaves, was led by flute players and followed by a weeping crowd to the pyre.

At that time it was by no means rare to find tame jackdaws or jays in patrician homes. They greeted guests with a cheerful "Good day!" and bid farewell with a regretful "Goodbye".

What might be termed the "exceptional" talking birds (the robin, starling, partridge and nightingale) were also highly prized by the Ancients. We are told of a very chatty white nightingale which was bought by Agrippa, mother of Nero, for the sum of 6,000 sesterces.

Parrots were not altogether unknown; they merely cost more. Varro speaks of them at length. From him we learn the etymology of the word "psittacine" (relating to parrots), which is simply a slight corruption of the original Greek word for bird – *sittake*.

Little by little parrots became fashionable. They were trained to flatter the rich and the great. They ate at their masters' tables and poets paid them homage. Ovid sang the praises of a green parrot whose statue was commissioned by Corinna, the Greek lyric poetess, after its death. Centuries passed, and the talking bird became a household pet and friend. Louis XII's cheeky parrots deserve mention as well as the one belonging to the Dames de Nevers, whose story is true. Because it sang the Credo so beautifully, the bird was sent to a convent in the South of France in the hope that it would amuse one of the nuns, who was dangerously ill. Unfortunately, during its journey down river to the south, the parrot came under the influence of bargees and forgot its sacred past. By the time it reached the convent, its vocabulary was so wanton and vulgar that it could not possibly be considered a suitable companion for a saintly woman. The parrot was returned to Nevers where it was obliged to do penance and relearn its Latin!

## Talking birds are like men

Many men prefer to express themselves in a vulgar fashion rather than in the polished style of orators or poets. So it is with birds. It is probably the musical sound of a word – its consonance – which appeals to them. Some learn more quickly than others. Some spend a lifetime repeating the same word whereas others acquire a whole sentence within minutes. This is possibly due to the anatomical formation of the mouth cavity or throat and perhaps it explains why all myna birds talk "without an accent" and parrots always speak "through their noses".

There may be other reasons, too. For example, just as we find human beings who are natural orators, so we find among talking birds some which possess greater aptitude than others.

Maurice Burton relates proudly and affectionately how his tame jay, Gaspard, in two years, learnt to imitate more than fifty different sounds. The most amusing was its imitation of the son of the house, whose voice was then breaking. For some time the jay confused the husky sounds of the boy and the little squeaks of his pet guinea pig. Gaspard certainly showed signs of intelligence, making associations which were not always coincidental and never failing to recognise persons he had met only once.

Finally, a word about talking canaries. A decade or so ago two canaries in south-west France achieved fame, having learnt to speak under amusing circumstances. Their mistress possessed two parrots to which she repeated the same phrase ten times daily. One morning she was overjoyed to learn from her young son that her efforts had succeeded. But the parrots were still dumb and it was the canaries in the adjoining cage which had started to talk!

This may seem astonishing, but in times past such examples were not exceptional. In the eighteenth century, a flautist from Nancy invented a flute called a Serinetto with which he serenaded canaries to teach them popular tunes.

## Do they know that they are talking?

Whatever the species, the same question arises: do they know that they are talking? Or is their nervous system and vocal apparatus simply like a record or a magnetic tape? By both these methods the recorded sound can be played back straight away, yet neither could be called anything but mechanical.

The secret of talking birds is more complex. Their nerve centres, stimulated by sound, record noises. Then orders are transmitted either to the various vocal organs (larynx, tongue, etc), to let the utterance through, or to the memory centres, to store the information. Exactly how and why this imitation takes place is just not clear even to the most eminent modern scientists. Hence so many different interpretations.

We must always remember that birds imitate spoken language only when in captivity. This brings us to a well-known social and emotional phenomenon, the fact that every young animal, living in close proximity to man, isolated from its own kind, "transfers" to man the feelings it would normally have for its own kind. It depends wholly on man and constantly requires his presence. Should the dominant figure leave it alone for a moment or a few hours, the animal grows anxious. Its whole being craves for this absent figure.

The bird psychologist, however, sees in this behaviour a similarity to that of a child separated from its mother. When alone and unhappy, the child instinctively cries, and its cries gradually take on the tone of the mother's exclamations and words of reassurance to the child on her return. This may perhaps suggest lines of research into the vocal reactions of both parrots and myna birds which depend on their excellent aural memory but at the same time are restricted by mere associations of feeling.

In the child, intelligence is awakened with the first spoken words, but language and intelligence go hand in hand, progressing through the association of ideas developing independently of associations of feelings.

Judging by the obvious ease with which young talking birds develop their vocabulary if they are taught early, certain animal lovers remain quite convinced that birds will eventually converse with human beings and understand them. This is an excusable mistake since appearances are often misleading. We can, therefore, forgive the reader who has been misled by stories. It is far more important to quote the expert opinion of modern animal psychologists and naturalists on this subject: "I am not a specialist in animal psychology", writes the great biologist, Jean Rostand, of the Académie Française, "but I can tell you that, to the best of my knowledge, there is so far no positive evidence to show that talking birds, and in particular parrots, possess true language involving appropriate usage of acquired words and communication with man."

Professor Clément Bressou, a member of the Académie de Sciences and the Académie de Médecine, does not think that the ability of certain birds to reproduce a few words or phrases can be likened to the ability to speak a language. "Their aptitude is developed artificially and is only acquired by living in close association with man and by methodical memory training. . . ."

And this trained zoologist, a former Director of the Veterinary School of Alfort, continues: "Furthermore it is not by articulate speech that parrots and other talking birds communicate with each other, but by soft twittering, far removed from the forced guttural sounds which they use when imitating the human voice."

"The ability to reproduce the sounds of human speech has no connection with language", asserts Professor Jacques Nouvel, Director of the Zoological Gardens in Paris. "It is more comparable to mechanical sound recording. Certain opposing views have been put forward when the word pronounced by the bird has coincided with a certain situation and environment. When a talking bird hears someone knock at the door, it might answer 'Come in' (which is perfectly explained by what we know of conditioned reflexes), but there are no examples of a conversation being carried on after this amusing invitation."

Colonel Philippe Milon, President of the League for the Protection of Birds and a keen promoter of interest in all species, has no hesitation in writing the following: "There is no real intelligent conversation between talking birds and their trainers. The 'word-sounds' which they make in no way resemble articulate language, the prerogative of man. Parrots, which I have long studied, both in captivity and in their wild state, communicate with each other by signal cries like all birds, using onomatopoeic sounds freely. Mocking birds can quite spontaneously imitate sounds which suddenly strike them, though the cause of their preferences is not known.

"Near La Ferté-Allais I knew a jay, completely

free and wild, which seemed to be an exception to the general rule concerning the need for training. It could imitate, well enough to fool you, the creaking noise of a certain cart which regularly followed a woodland path through its territory. . . . I also had an amazonian parrot, with a blue forehead, which imitated the noise of the running kitchen tap. This noise was certainly associated in its mind with the idea (? . . .) of a bath. Don't misunderstand me. I don't mean he was asking for a bath, but because of his behaviour, I am sure he associated the sound of the water with the sensation of a bath. There seems no doubt that birds can associate certain of their cries with relative situations or sensations; but talking birds make so many other sounds which have no sense or even opposite meaning, that it is impossible to speak of either reasoning or conversation. . . ."

Colonel Milon ends by proposing the following original idea: "It is perhaps not impossible that one day a species other than man will be able to make use of articulate language. After all, our far distant ancestors must have passed through a stage of 'signal-cries' and onomatopoeic noises, and the 'goo-goo' and 'ba-ba' noises made by very little children are nothing but that." Obviously this is what Rémy Chauvin calls "a somewhat romanticised notion of science".

Even so, it will no doubt be millions of years before there is any modification in the brains of birds. The truth is that every man sees himself as an Orpheus able to charm the beasts. We flatter ourselves to think that talking birds have a feeling of regret that they cannot establish intelligent communication between themselves and man.

It is for this reason that so many misunderstandings arise, as this striking example demonstrates.

One day a gifted and talkative parrot escaped from its cage. For several days it was sought in vain, then one evening it reappeared limping and emaciated and crying at the top of its voice: "I was caught in a ruddy trap!" The parrot could say nothing more, for it had forgotten all its extensive repertoire. As the bird repeated these words incessantly, its owner beamed with pride. "You see", he said, "my parrot talks! He is telling us all about his accident!"

Enquiries soon cleared up the mystery. A little boy, discovering the parrot with its foot caught in a sparrow-trap, had proudly taken the bird back to the village. When questioned he could only say: "It was caught in a ruddy trap!" (In French, "On l'a pris dans un foutu piège.") The terrified Polly could only repeat this sentence again and again and it was easy to hear it as "On m'a pris dans un foutu piège" (I was caught in a ruddy trap). No more was needed to establish striking proof of "intelligent language", in what was merely an approximation.

Lorenz, meanwhile, maintains that his tame raven could use one word intelligently – and only one. This was "roah", pronounced, Lorenz tell us, "with human intonation".

Such a statement, even from such a famous objective psychologist, is puzzling. The sound "roah" is the cry used by all ravens to warn their fellows of every form of danger. At most, it is the call which conveys instinctively its own feeling of alarm. Nothing proves, therefore, that Roah (the name Lorenz gave to his raven) was consciously sending his master a message by calling out "Danger!"

Professor Konrad Lorenz,[15] however, is far too

scrupulous to reach any general conclusion from this one example. "Not even the cleverest talking birds," he writes ". . . learn to make practical use of their powers, to achieve purposefully even the simplest object. Professor Koehler [one of the best bird trainers in the world] tried to teach the talented grey parrot 'Geier' to say 'food' when he was hungry and 'water' when he was dry. This attempt did not succeed, nor, so far, has it been achieved by anybody else." He then goes on to say. "The failure in itself is remarkable. Since, as we have seen, the bird is able to connect his sound utterances with certain occurrences, we should expect him, first of all, to connect them with a purpose; but this, surprisingly, he is unable to do."

It could not have been more neatly put.

## Notes

1 Joly, R. de — "A propos de pies", *La Vie des Bêtes*, (Jan, 1961)

2 Brosset, A. and Chapuis, C. — "Effets de la prédation sur l'évolution du cri des jeunes oiseaux", *La Terre et la Vie*, Muséum d'Histoire Naturelle (1969)

3 Chauvin, R. — *Animal Societies*, Gollancz (May, 1968)

4 Chauvin, R. — *Le comportement animal*, Masson (1969)

5 Frings, M. and H. — "The Language of Crows", *Scientific American* (November, 1959)

6 Dröscher, V. B. — *Le langage secret des animaux*, R. Laffont (1969)

7 Sire, M. — *Social Life of Animals*, Studio Vista (London, 1965)

8 Roedelberger, E. A. and Groschoff, V. — *Départ pour la vie* V.D.B. (Berne, 1968)

9 Bremond, J.-C. — "La Sémantique et les éléments vecteurs d'information dans les signaux acoustiques du rouge-gorge", *La Terre et la Vie* (June, 1968)

[10] Pellerin, P.     *Des oiseaux dans la vie*, Wesmaël-Charlier à Namur (Belgium, 1962)

[11] Grassé, P.-P.     *La Vie des Animaux*, Larousse (1970)

[12] Thorpe, W.     "Report on bird-song experiments", *Scientific American*, CCVIII (1963)

[13] Chiny, J.     *Les canaris*, Crépin-Leblond (Paris, 1959)

[14] Orth, H.     *Animaux de la nuit et du crépuscule*, Stock (1968)

[15] Lorenz, K.     *King Solomon's Ring*, Methuen (London, 1952)

# CHAPTER 7

# Lap dogs to elephants: the six thousand languages of mammals

So far we have discussed the question of language only in species which live in quite different environments from our own and which bear little resemblance to ourselves.

In the world of terrestrial mammals, we are treading on no less delicate ground. Even if mammals are closer to us than all other creatures, the roads and paths leading to them are none the less difficult to follow: often we cannot see the wood for the trees.

"They can almost speak." How often have I heard this phrase during my half-century of contact with pets! Nothing shows more clearly how little we know of the differences between the animal psyche and human intelligence.

Our beloved dogs and cats who live so close to us

often become distorted mirrors in which we catch glimpses of ourselves.

In my hunting days, I had as my gun dog and friend a Korthal griffon bitch named Miche. At a certain time each year her behaviour would change, to my amazement. Several days before the shooting season opened, a sort of intimacy was established between us which completely modified our habits. Our common way of life changed. This bitch, reduced by seven or eight months of city life to a sort of machine, eating, sleeping and wearily walking a few hundred yards daily, now became as one "possessed" and miraculously changed in character. She seemed to awaken from her state of utter lethargy, hardly eating or sleeping and, whatever the weather, rushing to go out provided I accompanied her.

At the end of the week, if some detail confirmed that our departure was imminent, she became literally crazy with impatience to set out.

Such behaviour will doubtless impress any animal loving hunter as commonplace. What I want to convey to the reader is something different – the degree of intensity with which Miche would raise her limpid eyes to meet mine when we were out hunting together. Better than any words could express, or without any other gesture, her "thoughts" were revealed perfectly. Her look established such an eloquent means of communication between us that one day, when I had twice missed an easy shot through simple tiredness, one of my fellow hunters was surprised to find me "discussing" my failure with my silently reproachful bitch. "All right, yes! I missed it, so what?"

My dear old Miche is dead now. She was devoted

A bee scout approaching a source of nectar. Von Frisch proved conclusively that a single bee can accurately transmit information on the position of a food-source by means of movement and mime.

*(Roger Gilroy – Keystone)*

The bee scout's mime varies according to the relative situation of the flowers to the hive. The circular flight, denoting a relatively close food-source, gives way to the more rapid and elliptical "tail-wagging" dance as the distance becomes greater. The angle formed by this figure with the vertical corresponds to the angle formed by the respective positions of the flowers, sun and hive.

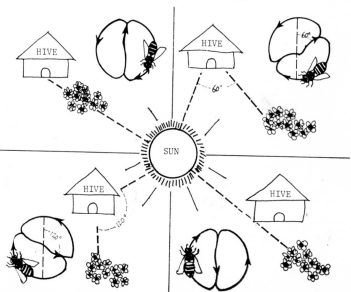

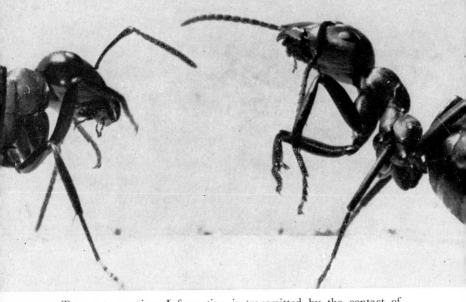

Two ants meeting. Information is transmitted by the contact of antennae.　　　　　　　　　　　　　　　　(*Baufle Jean-Marie*)

Ants feeding each other – a method of "mouth-to-mouth" communication.　　　　　　　　　　　　　　　　(*Baufle Jean-Marie*)

Common blue butterflies mating. The male can pick up the female's scent at a distance of more than a mile. *(Popperfoto)*

Crickets courting. The "love song" differs markedly from the "summons call". *(Jacques Six)*

A shoal of fish can present the image of one enormous fish when seen as a whole. The leaders form a spearhead followed by the wide middle section and the tapering tail-enders. This formation, like that of migrating birds, makes for ease of communication when a change of course is required. *(Serge de Sazo – Rapho)*

Communication between man and dolphin seems to be mainly visual, gestures being understood better than words. Hence the effects of surprising unison achieved by performing dolphins often in a matter of hours. *(Keystone)*

No satisfactory explanation has yet been given of the instinctive sympathy between dolphins and human beings. (*Keystone*)

Dolphins can travel under water at amazingly high speeds, guided solely by their unique "sonar" system. (*Keystone*)

A laboratory experiment in verbal communication. The submerged microphone transmits outside noises to the dolphin inside the tank. (*Keystone*)

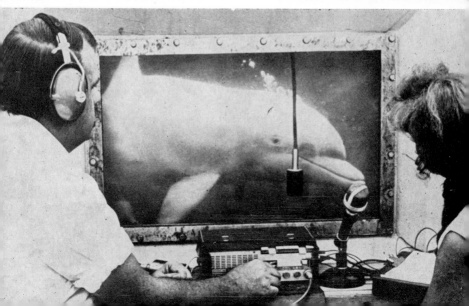

An unusual example of inter-species co-operation: a dolphin gives a seal a piggy-back ride in the Miami seaquarium. (*Keystone*)

A performing killer whale in the South of France. Despite the awesome size, terrifying teeth and murderous reputation of this mammal, Dr Griffin proved with "Namu" that an elementary form of communication with man is possible. (*Keystone*)

Two storks, with heads thrown back, "clack" their love-song. The same audio-visual signal is also used as a rallying call before migration. (*Jacana*)

Pelicans also give a rallying call before the great exodus begins. (*Keystone*)

Two Adélie penguins perform the courting ritual. (*Popperfoto*)

A flight of black-headed gulls. Like cranes, they express fatigue or hunger by uttering a sad moan before landing after a long journey. (*A. Molinier – Jacana*)

Each gull has its own distinctive tone of voice which its offspring recognise even before they are hatched. (*Popperfoto*)

Wild geese flying in the typical "V" formation. If the leader tires, it will be replaced immediately by another bird, but we still do not know how the new leader is chosen. (*Popperfoto*)

The courtship of the swan is graceful and dignified. After caressing cheek-to-cheek they entwine necks in a symbolic "love-knot" which binds them for life. (*Popperfoto*)

Bird-song, which seems to the human ear a random expression of joy, is basically functional, asserting identity and defining territory. (*A. Molinier – Jacana*)

The piercing call of young birds reaches their distant parents but appears to be beyond the range of human or predators' hearing. (*André Fatras – Jacana*)

Parrots are among the gentlest and most intelligent of talking birds, and have been kept as pets since earliest times. (*Keystone*)

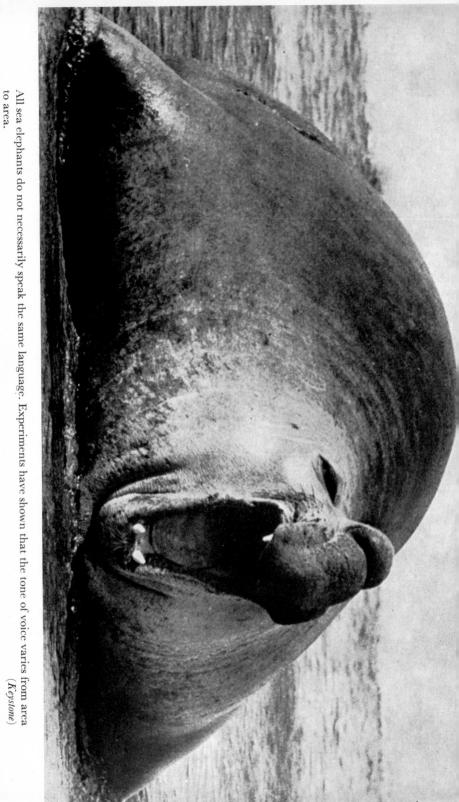

All sea elephants do not necessarily speak the same language. Experiments have shown that the tone of voice varies from area to area.

(*Keystone*)

These twin orang-outangs already have different ways of expressing the same feeling. (*Keystone*)

Wolf cubs engage in mock fights from an early age, preparing for adult combat which ends in a sign of submission from the weaker rather than in death. (*Popperfoto*)

The belling of deer at the start of the mating season echoes through the mountains.                                        (*Keystone*)

Rivals will be attacked if not driven away by the increasingly ferocious call of the dominant stag.                        (*Keystone*)

Marmots are always on the alert and signal the approach of danger by
whistling. *(Ian Tyas – Keystone)*

The lioness roars to scare away intruders but also to reassure her cubs.
*(Keystone)*

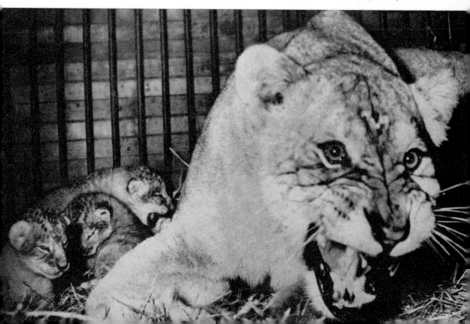

Horses are the most silent of the domestic animals, but their movements are always expressive. *(Yves Lanceau – Jacana)*

The immense range of both cries and moods sets the Siamese apart from other breeds of cat. *(Popperfoto)*

Being licked into shape ... The gentleness of the mother polar bear turns to ferocity when her cub needs protection. *(Keystone)*

Wolves have an instinctive sense of hierarchy and can indicate by the position of their heads a wide variety of social attitudes. *(Keystone)*

Male bears are aggressive defenders of their territory. *(Paul Senn – Popperfoto)*

The lowered trunk signifies the elephant's wish to intimidate.

*(Armelle Kerneis – Jacana)*

The trunk is raised when the threat prepares to become action.

*(Armelle Kerneis – Jacana)*

The chimpanzee seems capable of imitating a wide range of human behaviour, from simple facial gestures to more complex activities involving the use of tools and other hand-held implements. *(Keystone)*

The use of brain and hand gives the primates a superior advantage over the rest of the animal kingdom. *(Keystone)*

Baboons live in small family groups within communities. Much of their time is devoted to grooming one another. *(Popperfoto)*

In certain poses, the orang-outang looks strikingly human. Subject to violent outbursts, it is nevertheless calmer than the other primates. *(Popperfoto)*

This couple seems to be burdened with a domestic problem or a difference of opinion which needs careful handling! *(Keystone)*

The gorilla in captivity displays many facial gestures which resemble human expressions.
(*Zalewski – Rapho*)

Huskies, probably descended from wolves, are able to communicate in the most difficult conditions of polar expeditions.
(*United Press International*)

The few studies made of gorillas in the wild state reveal that oral communication is limited. Usage and response show extreme sensitivity to variations in the signals.

(*Popperfoto*)

Unmoved by the sporting event which their master is enjoying, these dogs reveal their state of mind by their very gaze: the large breed's solid composure, his small companion's trusting admiration.        (*A.F.P.*)

The author in relaxed mood. In his professional capacity he is President of the Académie Vétérinaire de France.

(*Photos Serafino*)

to me for nearly twenty years, answering no other voice but my own, acknowledging only my scent and presence. Never again have I found such close communion with any of the other dogs which have shared my life.

We must leave this question of strange emotional bonds. There is of course, no suggestion that such exchanges of "animal thought" and human response resemble an elementary form of dialogue. As Professor Paul Chauchard writes: "It is impossible to conduct a dialogue except between two beings both possessing brains sufficiently similar for them to understand each other."[1]

And yet, from all the evidence, my bitch understood me. There existed between us such an understanding and such close emotional ties that although she could not grasp the actual meaning of my words, she could both recognise my displeasure and, in her own fashion, recognise that there are limits to disrespectful behaviour which no well brought-up dog should exceed.

All "language", therefore, between ourselves and particularly sensitive animals cannot be totally dismissed. We make only one mistake: we forget that animals cannot interpret the actual words we use. When we talk to a dog, it hears a series of sounds, just as it would hear animal language. It remembers modulations, tones and emotive quality, and trains itself to react accordingly. It does not respond to the word itself, but to auditory stimuli ("signals" which are often reinforced by gestures or mime); all the rest is sympathetic magic ... like the myth of Orpheus.

## "Conversations" between animals and man?

We have seen that insects, fish and birds have precise hereditary codes which control the interrelations of each species.

For a long time, scientists made no attempt to understand these codes, probably because they seemed of no practical value in the exploitation of insects, fish and birds. But what of the "language" of mammals?

This important class, to which we ourselves belong, is made up of nearly 6,000 species divided into 19 orders and more than 160 families. The characteristics of this biological unit can be baffling when we consider how different in appearance are the tiny acrobatic female marmoset from South America, the placid lady elephant, and the bat, which suckles her young on the wing. Yet all these females belong to the same class. Each is covered with hair, each has a diaphragm which separates its pulmonary cavity from its thorax and – whether we are discussing a female marmoset, pachyderm (thick-skinned quadruped) or cheiropteran (mammal with membraned hands serving as wings) – all have identical mammary glands which provide their young with milk, which, although varying in quantity, is basically chemically similar.

Each of the six thousand species of mammal is furnished with a mechanism for communication and expression. Each has a larynx with vocal chords adapted to oral utterance, muscles and bony articulated limbs which make possible a "language" of gesture, and, most important, each possesses a brain which animates and controls the whole.

If we are to classify mammals, a few words must

be said about the brain, since science has shown an absolute parallel between the complexity of the brain and a creature's intelligence.

So far we have encountered only the rudimentary mental processes of insects (which are supplied with simple cerebro-ganglia) and fish (which are scarcely more complex) and batrachians (which despite their brains, are hardly better endowed than the higher invertebrates). In the case of birds, we have seen how controlled mental activity is possible, thanks to their superior form of brain. But with those species which suckle their young, we are immediately on a far higher plane.

In all mammals, the cortex of the brain already suggests that they possess intelligence, that it controls their actions and that they even have consciousness, as is demonstrated by certain facets of behaviour akin to "initiative". The brains of mammals are divided into two distinct and essential parts: the rhinencephalon (which controls the olfactory lobes) and the neocortex (seat of the more intellectual activities). The neocortex becomes progressively more complicated as we move from the lower mammals to the great apes.

The mammals mark the highest point of mental evolution in the animal kingdom. Beyond that is a barrier, through which no animal can pass. Although it may be obvious that in the more highly developed mammals the possibilities of communication are greater, even the most intelligent (and those closest to us) will never really be able to "converse" with man. This is not a question of teaching or training. Just as a motor-car (however powerful its engine) can never fly like an aeroplane, so the inability of an animal to express itself verbally is implicit in the construction of its brain. Between man and animal there is an unbridgeable chasm. An animal cannot talk. The real organ of spoken language is the brain, and the important thing is not its size but the organisation of its parts. As Professor Leroi-Gourhan points out, the very large brain of a monkey, equivalent in size to that of man's, can function only as the brain of a monkey.

An animal's brain, even that of the most highly developed mammal, is simply four times poorer in nerve-cells than the human brain and does not have sufficiently powerful motor nerves to allow it to modulate sounds. It is incorrect, therefore, to attribute the inability to speak to a difference in palate, tongue or vocal chords. It is equally wrong to cite the example of the parrot which can certainly imitate spoken language, but which is merely producing an echo with no "thought" behind it.

The precise difference has been clearly put by Professor Paul Chauchard: "Before becoming a means of thinking, human language is a means of communicating thought which pre-dates language. If language were merely a 'means of communication', man, in relation to other mammals, would merely have superior elocution."

## The nature of mammal languages

All language is social, allowing creatures of the same species to make contact with each other, utter sounds, understand signs, and reply. Whether instinctive or acquired, these signals or signs are designed to give information. It is in this sense that this book has used, and will continue to use, the term "language", except, of course, for real human language.

One of a herd of antelopes or zebras drinking at a water hole will notice the approach of a predator. Immediately its scent is picked up by the sentinel, the whole herd will take off at a gallop. But how are these warning signals conveyed? They are neither thought-transmissions nor given orders, but they are nevertheless communicated to the whole herd.

The French Ecole du Comportement et la Psychologie Expérimentale was the first to shed light on this problem, which for a long time had been observed more or less empirically and freely interpreted by means of analogy.

Today, with modern methods of investigation, the nature and significance of these various stimuli-signals are being established as closely and accurately as possible. But to obtain results, it was essential first of all to understand these signals; that is to say, to evaluate their sensory powers, range and strength. It was then necessary to test them, to reproduce them under the most stringent conditions, and submit an animal to their effects.

When we realise that mammals use, as means of communication, voice, scent, mime and gestures, and that in order to receive these emissions they sometimes use senses infinitely more perceptive than our own, we see that research in this field can be accomplished only with the greatest difficulty.

Let us attempt to describe these methods of communication, and first deal with sound. Rightly or wrongly, vocalisation is generally considered to be the most common and obvious of all the various forms of communication.

Communication by sound differs in certain basic

ways from other means of communication. Some animal species, for instance, are particularly sensitive to ultrasonic sounds. Following the American scientist von Békésy (winner of the Nobel Prize in 1961), who was the first to demonstrate that the inner ear works like a microphone, Buytendijk stressed the value of sound in communication, since it is equally effective by day or night and can penetrate more solid substances than can light or scent. Sounds can penetrate to underground lairs and burrows inaccessible to either scent or light. The transmission of sound signals is always under the animal's control and the strength can be varied almost infinitely. Sound is, indeed, the only signal which both transmitter and recipient perceive in the same way. Simultaneously with Buytendijk's experiment, the French scientist Dr Busnel and his team at the Centre National de l'Acoustique have, for the last fifteen years, been carrying out thorough research into sound, to learn more about wild animal noises. This is an immense task which can be achieved only by studying one species after another and by working in teams over a long period of time.

Consequently, many unsolved questions still remain to be answered by these dedicated research workers. For example, a hinny (the offspring of a stallion and a she-ass) whinnies or neighs, while a mule (the offspring of a donkey and a mare) brays. They both have similar genes, but in both cases it is the voice of the male which is predominant. Why?

We have no intention here of trying to deal with the "language" of all mammals, but only of those few (both domesticated and wild) whose modes of self-expression have become familiar to us through direct contact or observation.

## Notes

[1] Chauchard, P.    *Des Animaux à l'homme*, P.U.F. (1961)

# CHAPTER 8
# Wild mammals

Few works have been published on the sound signals of wild animals. It must have required infinite patience and scientific curiosity, not to mention physical endurance, on the part of B. Le Bœuf and Richard Peterson, to record throughout a whole winter the calls of sea elephants which always congregate off the shores of California at the beginning of December. These two American scientists noticed with surprise that newcomers to the island seemed to be unable to "speak" the language of the permanent residents. When the first recorded sounds were relayed to the native elephants there was no reaction. It was only after a few weeks, when the pitch of the newcomers' voices had been modified, and the "language" of these "foreigners" came to resemble the dialect of the native animals, that they were accepted. From now on they spoke with "the local accent".

Today, in Belgium, Doctor Nico van der Merwe of the Parcs Nationaux is conducting research into lions' language. Of all the great cat family, lions have the greatest range of vocal calls. At the moment, we know only that lions roar (particularly at night) and that they sometimes growl during rain or fog as if complaining at having to put up with such filthy weather!

In the course of his research Doctor van de Merwe has discovered a practical means of making the work of the keepers of the great national game reserves and parks easier. They must learn "lion language", as we would have to learn any foreign language in order to be understood abroad.

Meanwhile, the documentation we possess on the various sounds made by the inhabitants of jungle, mountain and forest is very restricted. Two African animals, originally of the same family, the giraffe and the okapi, were long thought to be mute. The fact is that both possess vocal chords, but, like horses, seldom give voice. A giraffe or okapi has to be very emotionally disturbed for the former to make even a low "moo" or the latter to utter a sort of neigh. And these sounds are not really signals. Giraffes and okapis rarely communicate with their own species other than by gesture or mime.

In contrast, most other African mammals are far from silent. The great anthropoid apes have a whole series of cries, shouts, gibberings and groans with which to express their feelings. In addition to these vocal utterances, they gesticulate in ways peculiar to their species. It would be interesting to know why the gorilla, for example, beats out with clenched fists a sort of tom-tom on his vast drum-like chest. Is it from euphoria, self-satisfaction or as a war drum to impress his enemies? There are many different explanations, but, to date, no experiment has provided a definite answer. What is certain is that the gorilla, really a friendly and gentle creature, only makes this noise when adopting a terrifying pose, standing almost upright and displaying his enormous hairy arms.

We encounter just as many difficulties if we attempt to make an exact inventory of all the different utterances of chimpanzees. In their native forests of central Africa, chimpanzees live such an organised social life that we readily see in them a certain regard for others and evidence of real traditions in some aspects of their playfulness. Not only have chimpanzees thirty-two different forms of cries, yelps and other utterances which they instinctively use under certain conditions, but many writers have given detailed accounts of a strange game played by these astonishing anthropoids when gathered in large bands.

"They construct", writes Garnes, "a drum made of clay which they place on a bed of turf which, being very porous, acts like a sound-box and amplifies sound. When the drum is quite dry, the chimpanzees gather round at night in large numbers and begin their merry-making. One of them beats violently on the dry clay, while the others prance around in grotesque and wild abandon. This saraband continues for hours."

This saraband, which the natives call "the monkey dance", is no figment of the imagination, having been recorded by many reliable witnesses. In the Budongo Forest, Frances and Vernon Reynolds watched this noisy spectacle seven or eight times and have given the following description of it:[1] "Calls were coming from all directions at once, and all concerned seemed to be moving about rapidly. As we orientated toward the source of one outburst, another came from another direction. Stamping and fast-running feet were heard sometimes behind, sometimes in front, and howling outbursts and prolonged rolls of drums (as many as 13 rapid beats) shaking the ground surprised us every few yards. Conditions were impossible for observation and after a while the most intense source of the noise moved off southward."

We still do not know the exact meaning of this display. It has been suggested that it is a sign of total hostility or intimidation, or perhaps it is simply a specific form of collective delight in noise. We know for a fact that monkeys in captivity derive particular satisfaction from beating the doors and walls of their habitat with the flat of their hands, the louder the noise the better.

Other monkeys living in bands, like the dog-faced baboons of Abyssinia and the Sudan, never stop addressing each other, indulging in such a wide range of shouts that it is hard to consider them any more significant than the various shouts of children in a school playground.

Yet certain sound signals made by baboons do seem to resemble a code. Males, posted as sentinels some distance away from the rest of the group, always give the alarm by the same short sharp cry. If the danger materialises, the strongest of the adult males echo these cries by barking like dogs and assuming defensive positions. The heads of families then shout precise orders to the females to take cover. Obeying these shouts – as emphatic as any human command – the females gather up their young and depart either calmly or hastily in single file and in a strictly determined order.

## European wildlife

The sounds made by European wild mammals include the growling of an angry old boar, the yelping of a fox, the almost smothered grumbling of a wild sow and the little distress cries of her young.

A summary of the sounds made by members of the deer family during the mating season will give some idea of the variety of their vocabulary, well known to huntsmen and specialists.

In autumn, at the first light of dawn at the edge of Alpine forests, the penetrating Ouaah ... Uaah ... Uaah of the seven-year-old stag (one with ten tines on its antlers) can be heard, answered by the deer on the next mountain. In the afternoon, the call becomes a long bellow: Ouaah ... Uaah ... Uaah ... Och! Och! ... Och! like the deep notes of an organ, making the air vibrate with echoes. A belling or troating (the sharper note of a young male) takes up the refrain. The lord of the mountain answers him with a raucous and menacing Och! Och! Och! which his young rival takes up, trying to emulate the same note. The adult stag takes offence at this, and, raising his muzzle to the skies, bellows even louder: Och! Och! The meaning of this exchange becomes clear. It will continue with increasing volume until the King of the Forest attacks or the younger rival prudently withdraws.

Simone Jacquemard has made excellent observations of foxes and badgers kept in semi-liberty on her estate, and has also given us valuable data concerning the sound language of roe deer.

As soon as the rutting season begins, the gentle elegant buck becomes an angry male, whose sharp, dry repeated barking announces that he is not interested in flirting. The females, or does, know that the alert has been sounded. Their agitation is betrayed by their startled eyes and loud, explosive hiccups, which are signs of unease.

Later, after the fawns have been born, and the mother, licking her delicate offspring here and there, takes them for their first walks in the father's territory, she finds herself faced with a really ferocious buck. This beautiful deer, which at any other time is all gentleness, can roar with rage more terrifyingly than a tiger. Does he mean to defend his young and impress his enemies, or are his roars merely a way of warning the mother that danger threatens? He shakes with furious spasms and then his terrible roars become sharp, regular low-pitched sounds. A few months later, these roars are replaced by a reassuring caressing, shushing sound, by which he now addresses his future mate. Trembling and crazy she dances about with all sorts of zigzag movements, whether from a desire to delay mating, or from an instinctive fear of the deadly arsenal which the aroused male carries on his forehead, we do not know.

In the mountains, the unexpected but unmistakable whistling of the marmot has also been studied closely. This has led us to a better understanding of the sound signals to which not only marmots but also their immediate neighbours, the chamois, respond.

Chamois, in fact, can make quite a similar noise. In both species, these reciprocal whistles, whether intentional or not, act as warning signals. Hardly has the little marmot, oddly seated on its haunches, uttered its warning call two or three times, than the alerted chamois will bound from ledge to ledge, while all the neighbouring marmots scurry to their holes. At the beginning of spring, the marmots take turns

in acting as "lookout" for their own species, signalling with a whistle the presence of an unwanted
creature within a range of 250 to 300 yards. Two
months later, at the beginning of summer, these
same sentries, doubtless more accustomed now to the
presence of strangers, only send out alarm signals
when man or beast is within about 100 yards.
According to the whistle's pitch and strength, maintains Bernard Clos, the attitude of the colony varies
from near-indifference through a passing feeling of
surprise to immediate flight.

Baby marmots are almost without fear. In July
it is not unusual to be able to photograph them a
few yards away if their mother is absent from the
burrow.

Beavers also appoint lookouts, but here the warning signal, though equally loud, is not made vocally.
At the slightest sign of danger, the "sentry" beats
the water with its heavy bat-shaped tail. Hearing this
noise, the whole colony dives under water and
disappears.

Rabbits in underground colonies beat on the
ground rapidly with their feet, to warn their fellows
of danger, as well as to express desire or anger. On
winter nights, when the ground is frozen hard, these
sounds can be incredibly distinct, carrying several
hundred yards through the silence.

Hares, which do not live underground and are
usually found alone or in pairs, do not give alarm
signals. Threatened with danger, they just run away.

The peculiar cry of the fox will be familiar to
some: a sharp bark followed by a piercing yelp. But
we do not know why adult foxes bark especially
before a storm or on wild and windy days. When the
vixen is on heat, she is followed constantly by a silent

panting escort of foxes, as bitches are pursued by dogs.

Paradoxically, little is known about the vocal language of our delightful European squirrel. More is known of the Californian squirrel, among whom Fitsch has identified five different tones of the same alarm cry, each signifying a different type of intruder or enemy – the snake, the low-flying bird of prey, the marten, the wild cat and man.

Thus we see the variety of sound signals made by animals in their natural state. It is possible, as Professor Bourlière believes, that the range increases as the sense of smell becomes poorer. Life in captivity, in fact, blunts an animal's sensory powers, but there are other factors, this time psychological, which affect the frequency and variety of oral utterances.

Naturally, in their wild state carnivorous animals remain silent, since the slightest sound can betray their presence. Once tamed, or brought to a zoo, these same species allow themselves to make their own particular sounds more freely. But are their yawns, roars and growls identical with those that would be uttered in the wild state? A comparative study of the vocal language of wild animals both in their natural surroundings, and under the influence of man, has not yet been made. Who will have the curiosity to try it?

## Notes

[1] Morris, R. and    *Men and Apes*, Hutchinson (London, 1966)
Morris, D.

# CHAPTER 9
# Domesticated mammals

The result of domesticating animals has not only been to prolong their lives. By providing them with food, shelter and protection, we have also delivered them from fear, and in doing so we have blunted their aggressive tendencies and reduced their panic reflexes to a minimum.

In most cases, therefore, animals dependent on man "talk" much more than they would ever dare to do in their natural state. The only exception is the horse.

The case of the domesticated horse deserves special attention. Horses make very few noises, apart from the whinnying of brood mares, which give prolonged "rallying" calls if their foals are taken away, or the joyful cries of gregarious young horses when put with a herd of their elders.

A horse never gives cries of pain, and is reticent in expressing pleasure, revolt or hatred. Sometimes, at the hands of a brutal carter or stable lad, it may suddenly show its anger. This takes the form of short, sharp resonant sounds, full of exasperation, accompanied by a rapid change in facial expression (lips curled, teeth bared, ears laid back) which all horsemen recognise as a preliminary to biting or kicking.

And this seems to be the extent of the need felt by "man's finest conquest" for oral communication with its conqueror. On the other hand, all the horse family are extremely sensitive to the human voice.

"Always speak to your horse!", young riders are taught, for, even when completely under our control, horses are always nervous, restless creatures. Thousands of years of obedience and servitude have not succeeded in reassuring them. And so it is necessary to use a restricted vocabulary (the same words always for the same action required of them) and to speak extremely gently.

In this respect, J.-Y. Delacour, one of today's most expert trainers, suggests quite seriously that different languages should be used to make contact easier between horse and master. For dressage he advises German because the short, guttural sounds, quickly heard and registered, are suitable for commands, whereas the softer harmonies of Italian are more suitable to calm, flatter or encourage.

Those particularly interested in the question of vocal exchanges between horse and master will gain much from the chapter devoted to this subject in Maurice Hontag's book, technically one of the best works of its kind.[1]


## The noisy Siamese cat

Cats have no hesitation in expressing what they feel or desire, either among themselves or to us. They growl, spit and utter strident cries of terror and anger. Their miaowing expresses equally plainly an urgent desire for food or freedom. Cats also purr, and we still do not know exactly why. Sometimes, after rousing themselves from a catnap they will make friendly little noises before risking a timid caressing lick.

Half-domesticated and half-independent, torn between the idolatry of ancient Egypt and the cruelties of the Middle Ages, the cat always remains on the alert and is always reserved, even in showing affection. Perhaps, like dogs, the cat is now becoming less wary of man and is prepared to be more trusting. But apart from the sounds common to this species as a whole, it can generally be said that tomcats rarely "speak".

Of all the feline race, one breed stands out in its ready self-expression: this is one of the most ancient breeds, the several thousand-year-old Siamese. During the first eight weeks of its life, a Siamese kitten keeps a prudent cat-like silence. Then after it has been weaned, it produces a whole range of cries, becoming increasingly frequent and varied as it grows older. There are cries like those of a new-born child, roars like a young lion and the low, velvety notes of a fine singer. No other cat addresses us with such insistence, making such continual demands, that we almost suspect an "intention" behind them.

The vocabulary of all female cats on heat is no less peculiar. While in most animal species it is the male which behaves in an extravagant fashion, the prize for the most piercing oral display of feline desire must be awarded to the female. Yet we do not know what causes her noisy madness, which rises from a gentle supplication to the most insistent howling.

Several of my own works[2] have studied the problem, but without resolving it. Yet perhaps a simple observation could help to direct research on this subject. Why is it that an amorous female cat howls literally all night long under the February moon? And why is it often sufficient to shut a cat in a dark cupboard to silence it, or at least to minimise its howling?

Can it be due to some central reaction involving vision and the endocrine glands, linked to the stimulation of nerve-centres by ultra-violet light on icy winter nights? Or, once the reflex action has begun, is it a sort of relay system of general alert, kept going by the cries and miaows of the initiator herself? This last possibility cannot be dismissed, when we remember the importance of sounds in stimulating sexual instinct, and particularly in awakening the maternal instinct.

In this connection, consider the following significant case. A pet Siamese, the self-appointed nurse of a new-born child, took her duties so seriously that she worried about the baby's feeding time and judged for herself the sincerity of the baby's demands. When the child uttered a cry, the cat would rush to the cradle then alert her mistress, remaining beside her, nose in the air, until she had finished feeding the baby. In the mother's absence, the cat would leap on to the cradle and lie beside the child.

This happens quite frequently and has been too readily attributed to jealousy (especially if an accident should occur, such as the child being smothered) or to the delight cats experience in sharing the gentle warmth of a baby.

Neither theory is correct. In the above example, the cats behaviour was so authoritative that it could have been motivated only by a form of "transference" to the new-born human child who had arrived in the house at an opportune moment (for example, at the time of a false lactation, or a "false pregnancy" in the cat). It is predominantly through sounds rather than through any other sense, that such chain reactions are set in motion. The wailing of the very young child arouses innate protective instincts in the cat; her maternal instincts then develop according to environment and circumstances.

The parents' readiness to attribute human behaviour to the cat explains the rest. However, it must be admitted that nothing so closely resembles the cry of a Siamese kitten as the wailing of a little human baby. We should not forget this.

## The dog – the silent suitor

Unlike cats, dogs do not give a vocal display of their sexual instincts. Quite the reverse. It is enough to see a bitch on heat, feverishly pursued by a number of panting, silent dogs, to realise that in these circumstances the "civilised" canines (much given to barking) revert to the prudent silence of the wolf. Now and again, without interrupting his pursuit, one of the interested parties may growl a rapid threat to an enterprising rival who has gone nearer to the female. The bitch herself "says" nothing, but continues on her crazy course without paying any attention to this conflict.

The mating of dogs is discreet and although many of their ancestral instincts may have been modified by their dependence on man, their sexual behaviour has not changed. It has remained exactly as it is today among wild dogs, which hate to be surprised in the awkward mating position.

It is therefore all the more surprising that, apart from the mating period, the domestic dog should be one of the most varied and articulate "speakers" of all mammals.

And we should not dismiss the possibility that a dog's bark may have been influenced to some degree by human speech. While hunting together, it is not impossible that man may consciously or instinctively have imitated the cries or barks of his invaluable partner to summon, guide or answer it and that the dog in turn may have echoed the onomatopoeia and exclamations of this superior two-legged animal accepted by it as the "dominant" of its species. The neolithic hunter, in fact, must have realised that the behaviour of the dog, which is swifter and better armed than himself, was in every respect comparable to his own. It would be natural for him to come to imitate it.

The belief that a perfect transmitter-receiver relationship in prehistoric times was responsible for the close ties between man and dog is in itself a fascinating theory. It is, however, no more than a theory, despite the mutual sympathy between the two species.

To attempt to show how, and to what extent, a dog reacts to the subtle nuances of the human voice, we must turn again to the studies and experiments of Buytendijk.[3]

In his soundproof laboratories, Buytendijk first taught a dog to remain on a chair, then to jump towards some appetising food placed on a table a few yards away when it heard a certain signal. A long leash, passed over a bar three yards above the chair, could either check the dog or allow it to run towards the food. Next, it was trained not to jump on the table when it heard a second sound two octaves higher than the first. After perfecting these initial experiments, the order and counter-order were modified by reducing the difference in pitch. When there was only a semi-tone between the two sounds, it seemed that the dog could distinguish the one from the other only with difficulty.

It was then decided to experiment with the human voice, the test being made on several dogs individually and under identical conditions. At the command "Jump!" or "Get down!" all the dogs hesitated before obeying. They seemed confused. Accustomed to mechanical sound signals, they were disturbed by the difference between these and the vocal orders.

The experiments were then made more complicated: either the dog's mistress gave the orders from behind a curtain; or the same orders were given by a concealed assistant who was unknown to the dog, or were given by the mistress, but recorded in another room and transmitted by loudspeaker to the room in which the dog was isolated.

In the first case, the dog obeyed without difficulty. In the second, when the voice was masculine and came from an unknown source, the dog hesitated for a moment, obeying the order to jump at the third repetition (having recognised the word?). The order "Get down!" was obeyed only after repeated hesitations during which the mistress gave encouragement.

In the third case, the dog cocked an inquisitive ear when he heard the faithful reproduction of his mistress's voice, but remained quite indifferent to the order. At every repetition, the dog turned questioningly towards the loudspeaker, and then, tired of not being able to understand it, he finished by lying down under the speaker.

This third and richly rewarding variation of the experiment made it clear that there is a marked difference between a mammal's reaction to mechanical sound effects and the response of less highly developed species. Where, as we have seen, crows unhesitatingly obey a particular "alarm call" which has been recorded and relayed by loudspeaker, dogs definitely recognise differences. Their response seems to be biological, depending on the presence (or absence) of a living "transmitter". Something similar distinguishes an actor on the cinema screen from an actor on the stage. In the theatre, there is contact and a direct link between spectator and actor. However, in the cinema, no matter how perfect the transmission, the actor only creates the illusion of his presence and there is no link.

Numerous experiments conducted in this field of research have confirmed the wide gap that exists between man's reaction to words (whose meaning he understands) and the reaction of animals to speech. All these experiments have shown that, even for a dog, words are only "signals" associated directly with some specific way of performing an action; the most intelligent dog is incapable of linking them to ideas.

## The perception of different languages

P. H. Scott James, who has studied dog's reactions to human speech, claims that out of three dogs which he taught to respond correctly to the meaning of words, two poodles acquired a vocabulary of forty words or exclamations; the third (a sealyham) was able to recognise almost ninety. Other observers have credited the dogs of really devoted owners with double that number.

We will not pursue with Scott James the astonishing theory that dogs possess a sense of time, and comprehend such abstract words as "Today", "Yesterday" and "Tomorrow". He is correct, however, when he claims that dogs can understand several languages, so long as this is seen as a simple matter of training (or self-training), responding identically to orders which sound different but demand the same reactions. There are plenty of examples to prove this.

One day I was visiting an English friend whose knowledge of French was such that he used it in the home as freely as his mother tongue. His pet chow was dozing at his feet while we chatted quietly. Suddenly the dog, which had ignored our conversation until then, raised its head at the precise moment my host was speaking, in French, about "greasing someone's palm (patte)". Hearing the word "patte", the dog was roused from its slumbers, got up and offered its paw to its master.

And dogs are not the only animals to adapt themselves to several languages.

In the region of Toulouse, four Friesian cows were found to be missing one morning. After several days of investigation, the thieves were discovered in an unusual way. The plaintiff, Mr Grossetti, who had lived in France for many years, was an Italian by birth and always addressed his animals in Piedmontese. Suspicion had fallen on two cowhands employed by a local cattle breeder, and it was suggested to Mr Grossetti that he should visit a communal pasture to try to pick out the stolen cows. On arrival he wandered among some sixty identical Friesians herded together. Then at random he shouted several orders in Italian, as was his practice. Four cows immediately left the herd and trotted briskly towards their real master. There was nothing left for the two thieves but to confess their guilt and give themselves up.

But to return to Scott James, how far are we to believe his assertion that dogs can perceive images relating to short sentences such as "Look at the aeroplane!" (whereupon the animal raises its head to the skies) or "Look at the train going by!" (when the dog looks around for a passing train).

This poses the delicate problem of animal thought, a subject with which the Ecole de Comportement is not greatly concerned, but which comparative neuro-physiologists are prepared to consider. "An animal cannot tell you what it thinks," writes Paul Chauchard, Director of the Hautes Etudes, "but neuro-physiological reflex action does not allow us to consider this aspect of man (endowed with interior thought) or animals (behaviour pure and simple) in isolation!"

Indeed, neuro-physiologists have shown that it is possible to control an animal's behaviour by applying electric impulses to its brain. Impressive as such experiments may seem to the outsider, this type of (almost painless) experiment has helped more than any other to prove the truth of the old proverb: "intelligence is born of the senses." It is the senses which register images in the brain.

The most intelligent dog is incapable of "evoking"

anything by itself, since to evoke is to think. The truth is that a dog's associative memory functions only under the influence of external stimuli – auditory, visual, olfactory, and so on. We then witness the release of a whole astonishing series of feelings and emotions. By making comparisons, we are tempted to give a flattering interpretation to this behaviour, but this is often dangerously wrong. There is no lack of examples. Here is one of the most classic.

Before going on holiday we entrust our dog to the care of strangers. Faced with its yelps and frantic behaviour, we find it difficult not to feel pity or remorse at the idea of abandoning our devoted dog to the "miseries" of separation. But every dog separated from its master forgets quickly. In a few days, it will become used to its new life. Mentally, it can no longer "think" of days past. But if in a month, or a year, a smell, a shape, or a noise should call up in its mind the image of the forgotten master, then we witness the delirious performance we know so well.

## From wolf to dog

Dogs, as a whole, have no scruples about making themselves heard, least of all guard dogs, sheepdogs and hunting dogs. However, like wolves, dingos, jackals and coyotes, there are some dogs which only give voice for a particular reason. These are nordic dogs, huskies and their close relations, chows.

Huskies cannot help barking with short, sharp impatient cries when being harnessed or when about to set off. They also respond loudly to their driver's words of encouragement, but once the team has set out only the leader transmits messages to the rest of the team on the long snowy journey.

Despite its promotion in the past hundred years to the rank of pet dog, the chow has remained profoundly influenced by its first masters, the silent Chinese. A chow will rarely bark, and it will utter only a few low, gutteral "woofs" to express pleasure on the return of its master. In the case of a fight, a brief roar is enough to roundly warn his adversary that combat is about to begin. There is usually a very quick struggle without bloodshed, even though its adversary may hardly keep to the ancestral code of wolves.

No breed is closer to the wolf than the Eskimo dog. The old saying, "Wolf doesn't eat wolf", which was previously mere observation, is now a scientific fact, confirmed by what we know of their behaviour. Because two wolves "speak the same language", and obey the same precise inherited code of ritual, the "submission signal" in theory inhibits the stronger one from pursuing his greater advantage.

I say "in theory" because there are exceptions. Jacques Marsault recalls the most famous case with amusement. During a lecture at Zurich on "Aggression", Konrad Lorenz had greatly interested his audience by revealing the truth concerning the generous behaviour of wolves at the moment of victory. Unfortunately, on the very same night, two wolves in the local zoo fought such a ferocious battle that one ended up with its throat savaged! The exception proves the rule. It is true that instinct is not infallible, and these were wolves in captivity, tamed and more or less changed by their contact with man. The social code may, therefore, not have over-ruled in the above case, though it would undoubtedly have done if the same two wolves had been in their natural environment.

Experts are now agreed that the domestic dog is descended from the wolf. Of the ninety behavioural characteristics present in the domestic dog, American research has shown clearly that seventy similar traits exist in wolves, and not only in captive, relatively adapted wolves but also among wild or semi-wild wolves in the great Canadian nature reserves. Jackals, coyotes and foxes, on the other hand, show very few of the characteristics of domestic dogs.

The great difference in vocal sounds made by dogs and wolves must therefore be attributed to domestication. For a long time it was believed that wolves howled but could not bark. Darwin was sufficiently convinced by this weak argument to formulate his theory of the dual origins of these two closely linked members of the dog family. However, although it is true that the wolves' usual noise is their well-known sinister howl, which is a rallying call, they make current use of two or three other signals very like barks when near their lairs.

When very young puppies are first separated from their mothers, they also howl plaintively, as if to say, "I'm all alone, I'm so lonely!" A few days later the howls become shorter and more pleading, interrupted by little demanding barks all the more frequent between the ages of a month and a half and two months, because the puppy has difficulty in correctly locating the sounds it hears. From now on, the barking grows progressively more demanding as the silence of night gives way to daylight and the puppy can once again take its place among the human pack.

All adult watch-dogs make the same short sharp

sounds as soon as anyone approaches their territory, then, according to the intruder's behaviour (and of course according to the dog's breed and training), a whole range of sounds will follow this initial alarm. We are now beginning to interpret these sounds. There are the smothered roars of simulated attack or retreat, modulated sounds of friendliness, the nervous barking of joy, sounds that signify fury, fear and grumpy withdrawal made to save face.

These differences between the sounds of wolves and dogs deserve closer attention.

Wolves and dogs have the same vocal apparatus. The length of their vocal chords is almost identical and all dogs of the lupine class have similar resonating systems, made up of the mouth, nasal cavities and roof of the palate. Lung capacity and the force of exhalation of dogs and wolves are also comparable. Wolves and dogs therefore ought to make the same rich variety of sounds, at the same pitch. But this is precisely one of the main points of difference, and the explanation for the complex vocabulary of the domestic dog seems to lie in the calming influence of thousands of years spent in human society. Professor Leroi-Gourhan virtually acknowledges this when he writes[4] that the evolution of dogs follows much the same pattern as that of man on their respective planes of quadruped and biped.

As far as the evolution of canine vocabulary is concerned, how can we avoid making instinctive comparisons with man? For example, the noisy yawn of boredom (which is a residue of the wolf's howl), the rising, strident clamour of dispute, the puppy's cries for help, the sharp notes expressing fear of being hit, the exasperating yaps of stupidity, the unconvincing warnings of old watch-dogs, the enraged cries of provocation, the snivels of loneliness, the low growls of anger ... All these sounds, from barks to gentle murmurs, with a hundred variations, are so obvious, familiar and expressive, that we have no need to translate them.

## From Saint Bernard to fox terrier

Leaving aside all question of imitation, how could dogs not vary greatly in the sounds they make, when they vary so greatly in shape and appearance from species to species? With such dissimilarities in the shape of the head, the height of the roof of the palate, and the size of the thorax, how could a greyhound and a bulldog have the same voice? For similar reasons, while all foxes yelp on the same note and all wolves howl in the same way, a fox terrier could never bring its shrill bark down to the low bass of a Newfoundland dog or a Saint Bernard. Nevertheless, these differences do not prevent all breeds from speaking "dog" and from understanding each other.

## Deliberate communication

Now we come to the delicate question, whether a dog is capable or not of consciously "collaborating" with its own kind.

Jean-Claude Filloux and numerous other animal psychologists think not. According to them, one must distinguish between a partner to whom one responds instinctively, as if to a stimulus, and a true "collaborator", a fellow of the same species, whom one is consciously able to help and who can in turn give help. In animals, this sort of communal effort is mentally impossible. Many specialists (Wolfe, for example, on the lower monkeys, Daniel on white rats, Crawford on chimpanzees) are quite definite on this subject. However highly developed an animal may be, it cannot recognise in another animal "another self", capable of co-operating with it in a communal task.

But then how can we explain the behaviour of the leader of a team of sledge dogs? This dog, commonly known as the Boss, is leader not only in the sense of "dominant" but also in the sense of guide. It is this dog, placed at the head of the team, which pulls the hardest and is the first to plunge waist-deep into the snow to clear a path. It also receives the sledge driver's orders. All this could be no more than the result of training if the leader merely had to set an example; but it does more than that. If one of the team slackens its efforts, the leader, looking back quickly over its shoulder, picks out the guilty party and addresses to it a couple of sharp barks. This is the "recall to duty" signal, showing that the leader is not to be fooled and knows full well which dog is not pulling its weight. The culprit has been warned – and rightly. On arrival at the next halt, no sooner have the dogs been unharnessed than the leader rushes up to the guilty party and gives it a good telling off, which is understood and always accepted meekly.

With all due respect to the objectivist school, how can we fail to see in such behaviour the possibility of intentional communication sought after for its own sake?

Take another more classic example. A dog barks at

a closed door. Animal psychologists maintain that it does not bark to be let out, but because it is shut in. The distinction is very subtle. Naturally we cannot help agreeing with the animal psychologists if we accept that "animal language" never expresses a thought, but is only an automatic signal. But if we prefer to see these signals as informative, in that they warn us of a vitally important situation, how can we fail to see anything but an abstract "idea" in the very notion that the dog regards itself as "shut in"? But there is also another very plausible explanation: the reaction of every dog confronted with something unusual is to bark; it is his way of alerting man, his superior. What better help could he have than this man, who will come running when he barks?

Does this show real understanding of the situation, a flash of intelligence, conditioned reflex, or self-training? Let us just say that in these circumstances, dog regards man as a tool.

It seems that dogs begin to communicate among themselves from an early age. They are blind and deaf and scarcely able to walk during the first twelve days of their lives, yet by whimpering almost from birth they keep in contact with the mother and are instinctively assured of food and warmth.

Young wolves and all other wild members of the dog family behave in the same way. Like puppies, newly born wolf cubs use their shoulders and muzzles and utter little whimpering sounds to make their first social contacts and establish among themselves a form of hierarchy in obtaining the best place at their mother's teats or to snuggle up against her. Thus begin their first exchanges and the integration into communal life. Then the mother-dog or vixen begins to make more complicated sounds and other signals as the young grow up. Both wolf cubs and puppies inherit the instincts of their species. But for dogs this is only a beginning. They then have to become familiar with a new environment and a new language, consisting of new gestures, words and smells. This is by no means the least complicated or subtle part of a dog's life, nor is it the least source of misunderstanding.

## Notes

[1] Hontag, M.    *Psychologie du cheval*, Payot (Paris, 1954)

[2] Méry, F.    *Le Chat, son énigme,* Pont Royal, R. Laffont (1968)
see also: *Just Cats*, Souvenir Press (London, 1957)

[3] Buytendijk, J.-J.    "Traité de psychologie animale", *Introduction aux études philosophiques* (Paris, 1952)

[4] Leroi-Gourhan, A.    *Le geste et la parole*, Albin-Michel (1969)

# CHAPTER 10

# Beyond vocabulary: the secrets of gesture and scent

If you want a horse to follow you, take it by the bridle, and with your back to it walk alongside. Above all, never look at it. Like monkeys and many other animals, horses dislike being the object of man's direct gaze for any length of time. This may be because they feel cornered and disconcerted by this means of communication (the real nature of which escapes them), and thus find relief from such profound uneasiness.

Here we touch on a mysterious aspect of what is mistakenly known as "silent language". It may be soundless, but is nevertheless eloquent– and more so among mammals than among any other class of animals.

## The sensitive horse

We have already stressed the limitations of the horse's vocabulary. But its mime, postures and gestures substantially compensate for this lack. With

141

this wary and emotional creature, all expression is spontaneous and direct.

If a horse is in pain, a vet needs only to look at it to know how serious the condition is. If the trouble is sharp intestinal pains, the horse will be restless, pawing the ground with its front hooves and turning its head towards its side. When it has had a stroke, it will make violent uncontrolled movements at complete variance with its dazed appearance, fixed eyes and dilated nostrils. If it falls to the ground in an abnormal position (like a sphinx, on its knees or like a seated dog) the diagnosis will be worse. And contorted lips (comparable to "sardonic laughter" in humans) are another serious sign, meaning often enough a hopeless case and a fatal outcome.

A healthy horse displays its condition just as clearly. It will paw the ground with impatience and tremble anxiously. If under constraint, it will lay back its ears, lash out, shy abruptly or quickly back away – all ways of showing its discontent.

When we see a large draught-horse, or any carthorse, labouring over a task, it is difficult to imagine how sensitive to touch a horse can be. Observe a saddle-horse. Riding, and the various contacts implicit in this sport, focus attention on the many unsuspected shades of meaning which are communicated by the touch of horse and rider.[1]

For the horse, sight, hearing and smell are certainly valuable means of communication with the surrounding world, but the sensitivity of its muscles and skin plays a far more important role than its other senses. Despite the still rough texture of its coat, a young foal can, from a very early age, send ripples over its skin to rid itself of annoying insects. The hairs in its nostrils are delicate antennae, as sensitive as our own fingertips. To prove that they are used in the same way, it is enough to cut off these hairs: we see the poor animal's lips bump into the edge of its trough as though it has been blinded. Nor must we forget to mention its four legs. These are so sensitive that it has been said, "a horse sees with its feet". Feet and legs constitute such an excellent radar system that fifty horses at full gallop will pass over a fallen rider without even touching him so long as he remains perfectly still and makes no abrupt movement.

These few brief examples give some idea of the horse's exceptional receptivity to stimuli communicated by the rider's touch.

It is not within the scope of this work to describe different equestrian techniques. I should merely emphasise that the whole art of equitation would be inconceivable without the "code of touch", by which the horseman communicates with his mount more effectively than by the use of words. This "dialogue" is established through the horseman's hands and legs, and through what has long been known as the "seat"; that is to say, different ways of distributing his weight to match the animal's movements.

It is agreed that horses rarely "talk" but, from the tip of their satin muzzles to the points of their sensitive feet, every part of their bodies vibrates like the strings of a harp. (This has been shown by Dr Roger, an army vet using acupuncture.)

## The cat: creature of extremes

Cats are just as 'tactile" and vibrant as horses, and have equally expressive languages of mimicry,

143

contact and gestures. Even the range of "feline utterances" seems unnecessary once we are familiar with the everyday behaviour of a young cat. A Siamese cat could get by without miaowing. In less than ten seconds its innocent sky-blue eyes can turn royal blue with fear or violet with anger. When surprised, its eyes open extraordinarily wide, but suspicion reduces them to narrow slits. A rapid almost imperceptible twitching of its jaws betrays the hunter about to pounce. Every slightest fold in the chocolate-coloured mask covering that pale face marks a change of mood.

Such are the principal signals used by cats to express their wishes and feelings, whether they originate from the Far East or from our own part of the world. We should also mention that they arch their backs, stretch their legs out behind them, roll over, turn somersaults, turn on their backs, and when on heat, twist about convulsively almost like snakes. And we should not forget their slow, silent tread and the rolling shoulders of fighting tom-cats, slaves of their lady-love before they are masters. There is also the curious behaviour adopted by the "dominant male" in the colony of cats towards a newcomer. If a strange cat comes into the group, the boss-cat will approach it straight away, and sniff it. An instant assessment is made, resulting in either a fight or (more often) friendship: the stranger has come in peace, and begs indulgence. The dominant cat then seizes the other by the scruff of its neck without really biting it and drops it or at least goes through the motions of throwing it out, then lets it go without a murmur. This gesture makes it quite clear who is master, of both cats and territory.

The aggresive or defensive attitude of cats when

144

fighting is also peculiar to the species. So much has been written on the subject that it is easy to confuse the different forms it can take. The reactions vary according to whether it is confronted by one of its own kind, by another species of animal, or by man.

Confronted by a defiant cat which is prepared to yield to neither threats nor acts of aggression, the "dominant" animal advances slowly until the combatants are face to face growling softly and waving their tails in the air. With bristling fur they remain almost motionless for a while, each looking for signs of hesitation or imminent action in the other. Then suddenly, spitting furiously, they spring into attack, gripping each other and lashing out with their hind legs. Each tries to wound the other to the quick until one breaks off the fight and slinks away.

Confronted by a threatening dog, a cat's gestures are different. Here we find the classic arched back, the rigid straight legs, the waving tail swollen to twice its normal size, the ears laid back, the nostrils and muzzle wrinkled and the mouth fully open, revealing fangs ready for action. The cat is, in fact, terrified. It is hoping that the dog, exasperated or surprised, will change its mind and make off without further ado. But can the dog escape? The cat's growling suddenly becomes a war-cry, and like a demon it catapults itself at the dog's head, clawing furiously at its eyes and nostrils. Then, taking advantage of its stupefied adversary, it makes off before the other has collected its wits.

Faced with man, a cat's attitude is again quite different. Generally speaking, it will avoid any form of aggression, taking flight at the slightest gesture. Should it be at a veterinary surgery or in an exhibition ring, where it cannot escape, it is no easy matter to catch hold of it against its will. According to its temperament, it will either give a pleading worried look, or grimace, with its eyes flashing in anger. But in either case, it will raise its paw, ready to strike. This is fair warning to keep your distance, a warning that should not be disregarded as it can be quickly followed by a lightning attack from its unsheathed claws.

## The dog: domesticated wolf cub

These brief sketches of the most common displays of feeling in horses and cats give an incomplete picture of the intensity of their feelings. The horse is too far removed from us emotionally, while the cat is too egotistic and secretive for us to speak of "shades of feeling" (to use Henri Piéron's phrase) or to think that its reactions really correspond to the degree of feeling they seem to express.

When we discuss dogs, the most socialised of mammals, in human terms, we are on even more delicate ground. First of all because in some areas a dog's sensory perception is so great that our own senses are too weak to evaluate it. There is no parallel between the causes of a given feeling in a human being, and its causes in a dog. For example, how can we determine the kind of stimulus that will make a dog suddenly leap at a passer-by? Is it smell, which we know to be such a vital source of stimulus-signals? A dog is so sensitive to oily acids that it can detect one billionth of a milligramme of butyric acid in a cubic centimetre of air.[2] In simpler language, this means that our sense of smell is a billion times weaker than that of the humblest mongrel! Is it

146

sound? Does the dog recognise the human being by the sound of his footsteps or by his voice, or perhaps by some other sound he makes? Any aural stimulus is valid, since a dog's ear is sensitive to ultrasonic waves and can register up to 90 or even 100,000 vibrations per second, while the human limit is barely a quarter of that.

This only goes to show how careful we must be, in the absence of all controlled experiment, in attributing qualitative significance to various expressions and attitudes of dogs. A low growl may equally well mean "Come any closer and I'll leap at you", or "I'm scared and would rather be elsewhere."

Since we cannot speak dog-language, how can we hope to know exactly what motivates it? We will therefore confine ourselves to methods (other than sound) by which dogs communicate among themselves or with man and which have been traced in wolves which are free of all human influence.

**From wolf to dog**

Like a little wolf cub, a puppy makes social contact with its own kind at a very early age. Puppies and wolf cubs alike, from the time they are weaned (and often even before), will engage in games which are sham fights. By playing at attack and defence, they learn the significance of mimicry, cries, postures and gestures.

At this point the puppy's play is so instinctive that it does not require the presence of its brothers and sisters to act as partners. From the age of a fortnight the early swaying movement of the head, used to locate the mother's teats by touch, is replaced by the similar movement of shaking objects from side to side symbolically, as if they were victims. A little later comes the need to tear at this prey and this time the symbolic victim is the weakest of the litter. Already this "grip" on the nearest living creature – apart from its mother – reveals the hierarchic instinct and the feeling of superiority. At this stage, the victor, just like the dominant cat, is happy merely to nibble the victim's neck, and again, like the cat, makes as if to dispose of it, but in play. Yet if this little bully is taken from its family and familiar surroundings and placed, say, in the middle of a room, it will try two or three times to escape but will soon stop and give the distress cry. This is equivalent to the howling of wolves, taken up and repeated by the whole pack to show that a lone wolf's signal has been recognised.

Presently comes the last phase of early "childhood" which leaves the mark of the environment on the behaviour of the wolf cub or puppy. As soon as a cub or puppy can see and hear, it attempts to trot around and stand on its hind legs. A few weeks later, it will be able to jump, scratch and gallop. So long as its games assume the more serious nature of real fighting (with enticements, threats and the growing wish to be the winner) the relationship to its environment will be cautious and peaceable. Its ears and tail and its facial expression take on precise meanings. The animal is teaching itself the signals proper to its own species. Schenkel, Meischner and a number of other Swiss and German authorities have drawn up quite a complete inventory of such behaviour.

Like wolves, dogs turn round before lying down because in their wild state, like many other animals, they are forced to assume a curled-up position to retain maximum warmth for their bodies.

Again like wolves, dogs which are sure of their own superiority have a calm look with heads held steadily but not stiffly and ears almost still. They express alertness and latent threat by raised brows, stiffened necks and wide-open eyes. Shifty sideways glances herald escape or flight, while a lowered head with shoulders pushed forward and forehead set firm denote cool courage and a readiness to fight if provoked.

From observations made over a number of years on wolves living in a semi-wild state, Schenkel drew up a list of head-positions, facial expressions and tail movements which seem equally applicable to dogs:

"Head held high, direct gaze, tail bent over like a curved blade, motionless ears – all denote the proud assurance of a leader.

"A wolf lower down in the hierarchy will draw in its neck and look down at the ground, lay back its ears and imperceptibly wag the tip of its tail.

"Neck extended, muzzle horizontal, tail moving independently, express a half-threat, usually made without conviction.

"Head thrust forward, ears flat to the skull, jaws open and gums bared to expose fangs in readiness: all this means attack.

"Jaws barely open, ears half-cocked, tail between the legs and the general position almost a crouch, denote that the subject is of lowest rank in the pack, and resemble the attitude of a submissive dog trembling at its master's voice and cowering even before it has been struck.

"There is one stage lower: the uncontrollable passing of urine – an appeal for mercy to a superior or undoubted victor. There is a strange and subtle difference between victorious wolves and our own victorious domesticated dog. The victorious wolf siezes its victim by the scruff of the neck and holds it powerless for a moment before releasing it. On the other hand, a dog knocks down its victim and holds it there with a firm paw, but it stops short of tragedy if the victim has the good sense to offer its throat to the victor's teeth."

These are the principal gestures found among dogs and wolves. From what is known of wolves, we have no reason to doubt the meaning of these signs.

## Scent and the language of smell

We know less about animal scents as means of communication. Although a cat may not depend on its sense of smell, it is often by the "language of scent" that it communicates with its own species. A she-cat which has been deceived will sniff the fur of her faithless husband to distinguish the odour of other she-cats before unsheathing her claws. A mother cat scrupulously checks the number of her young by smell and recovers any that have been removed by a jealous neighbour.

What meaning should we give to the behaviour of even the most "humanised" pet dog, which, meeting another dog, will extend its nose to the other's hind parts? We know that dogs do not sweat. Being particularly sensitive to biological odours (for instance, urine, blood and sweat) perhaps they have no better means of greeting or getting to know each other than by sniffing the very distinctive odour of secretions from their fellows' anal glands (or excreta).

I knew two English greyhounds which had had perfectly clean habits for ten years and even refrained from fouling their own garden. Suddenly, despite their training and all attempts at correction, these well brought-up dogs began daily to foul the furniture, rugs and every corner of their master's rooms. Now it so happened that eighteen months earlier their master had acquired a parrot. Surely, after eighteen months, the dogs' behaviour could hardly be due to jealousy. No, the explanation for their conduct was much simpler: one day, precisely eighteen months after the parrot's arrival, it had learnt to bark. The greyhounds, astonished to have this invisible, scentless, but perfectly audible dog in their midst, were warning the phantom that it was trespassing on their territory. Unable to meet the intruder, how else could the disconcerted greyhounds communicate with it?

Much has been written about dogs' olfactory code. Although we are beginning to understand the extreme subtlety of their sense of smell, many problems still remain unsolved in this field of research. How, for example, do tracking dogs follow the spoor of game? If we accept that a dog can distinguish a difference in intensity between an old scent (i.e., the direction from which a creature came) and the more recent (i.e., the direction in which it is going), we must conclude, as Aimé Michel rightly puts it, that dogs possess a sensitivity to smell which is difficult to relate to the chemistry of scents. We can only suppose that a dog is conscious not only of the scent of the individual it is tracking, but also of the scents which the individual has picked up from the terrain over which it has passed. Dr J. Chavance, of the Veterinary Academy of France,[3] postulates that in addition to the changing degrees of intensity, the very nature of the scent must also change as the trail progresses through stages marked by variations in molecular pattern.

This suggests something comparable to the intensity or frequency modulations which occur in radio waves. To prove this it would be necessary to find some method of interpreting the "scent trail", then find if our methods of registering scent correspond to those of dogs. To do this, we would have to make artificial trails, both continuous and broken, with increasing or decreasing intensity. Experiments along these lines were attempted some thirty-five years ago in Germany during an International Congress on Olfaction. As secretary-general of this congress I can confirm that these experiments did little to further our knowledge: the tracker dog's sense of smell is still a complete mystery.

Animal psychologists are becoming more aware of the important role played by smell in relations between mammals of the same species and of different species. The Americans have already shown that adult mammals respect their young and their newly born offspring, not because they are young, but because their mineral salt content (magnesium in particular) provides them with a personal odour which usually inhibits the adult from attacking them. A sow's maternal instinct is much more clear-sighted than that of many other domesticated mother animals. A bitch deprived of her own offspring will even go so far as to nurse chicks, although they are quite incapable of relieving her swollen teats, whereas a sow will only rarely consent to feed piglets of another litter (and then they must be brought to her within an hour of birth, so that they carry no other sow's odour).

Experiments conducted by Klopfer and Gambler on pregnant ewes deprived of their sense of smell, either totally or partially, by removal of their olfactory lobes, have given interesting results. Ewes which have not undergone the operation will always accept their own lambs, but will reject any stranger. All ewes (before or after giving birth) which have been subjected to the complete operation will accept any lamb shown to them, whether their own or not, without distinction of age, size or appearance. Thus, neither sight nor hearing plays any part in the recognition of their own offspring. As for the ewes partially deprived of their sense of smell, the maternal behaviour is confused to the extent that they will accept without distinction a certain percentage of their own lambs together with those of other ewes.[4]


## Smelling each other

As a result of similar operations on rats, the American neuro-psychiatrist Dr Harry Wienner, of the Metropolitan Hospital, New York, discovered the existence of ICM (Internal Chemical Messages). With certain animals, smells are of major importance in establishing relations between members of the same species as well as different species. The senses need only be slightly disturbed or more than usually receptive to produce the most unexpected reactions. For example, over half a century ago, Professor Cadéac, of the Veterinary School of Lyons, demonstrated that the smell of angelica produced hallucinations in hunting dogs. Responses such as this, provoked by smell, remain quite inexplicable.

For ten years I kept a racing greyhound, a truly hyper-sensitive dog, which displayed the most unusual symptoms. To pass a butcher's slab in the summer, or even the open door of the shop, gave this thoroughbred such an attack of catarrh that one could follow its tracks for the next ten minutes by the drops which fell from its nostrils.

Today Dr Wienner maintains that, just like man, animals are influenced by odours (especially biological ones) which can have the most unsuspected influence on the sexual behaviour of mammals.

If some thirty female rats are shut up together in a narrow cage, their sexual activity is completely upset. They show all the symptoms of "false pregnancies" and ovarian upsets, followed by disturbances of their maternal instincts. However, if a single male rat is brought among these half-crazy females their reproductive patterns quickly revert to normal.

Even more strange, if one of the females is pregnant, and a rat other than her mate remains beside her even for a quarter of an hour each day, the foetus ceases to develop, or else she suffers a miscarriage within a week. But the rat's presence is not necessary to induce this phenomenon; the smell of its urine is enough to produce the same result.

In the circumstances, it seems that the sense of smell is directly responsible: if the females' olfactory lobes are removed, leaving them with no sense of smell, they present no such irregularities.

# Some strange "signals"

Among humans, scents also play a far from negligible part in communication. We all know how chemically made perfumes, as well as natural fragrances, can affect us. However, our sense of smell is, comparatively speaking, so poor that we are inclined to doubt its value as a real form of "language" among so many other mammals. Yet, the facts seem to prove this. For example, martens mark out the whole length of their territory by secretions of the anal gland, also marking the ends of broken branches as boundary posts.

Among stink-martens like the polecat, weasel and stoat these secretions have a dual value serving also as a means of defence. These animals can project their liquid secretions at will, and to such good effect that American skunks (very aptly termed "mephitic" – evil-smelling) are able to repel the boldest predator by this chemical bombardment.

Beavers produce the well-known castoreum (a secretion from two glands near the reproductive organs). Like many other rodents, beavers use this scent as a silent warning to other species to keep their distance. Unfortunately, fur trappers can also use this "signal" to lure large numbers of beavers, finding it by far the most effective of baits.

In the chamois, the scent is secreted by glands behind the horns. The male makes use of this scent (particularly during the winter) to define the boundaries of the territory reserved for himself and two or three of his wives. This wise precaution prevents many a fight.

As for the other animals of our woods and forests, it is their urine which serves as a mark of their

155

presence, a boundary post or a secret weapon. The European bison, after tearing away part of the bark of several trees, empties its bladder a few yards away, rolls in the liquid (as well as it can, because of its hump) while it is still warm then rubs its head and body against the stripped tree trunks.[5]

The reindeer leaves deposits of secretion from the glands between its toes. Not only will other males turn away from such spots, but they will never pass by there again.

Both wild rabbits and squirrels sprinkle their mates with urine. This strange method of sealing their conjugal relationship, or blinding their rivals (sometimes from more than a yard away), is performed with incredible precision.

The hippopotamus leaves the water every evening to follow the same tracks leading to the interior. The narrow well-trodden paths are literally lined with pats of dung and sticky urine. In this way, each glistening hippo marks out its own territory which it stoutly defends. According to Hediger,[6] only the male hippo indulges in the following odd method of establishing his territorial rights. After smelling out the spot which he has carefully chosen, he generously loads his little tail with the necessary excreta, then rotates it rapidly so that solid and liquid particles are dispersed in all directions, impregnating the surrounding area with his particular scent, and establishing it as *his* territory.

Still on the subject of hippos, the Director of the Basel Zoo asserts that we are wrong to think of their wide yawns as a sign of boredom or indifference. On the contrary, this full opening of the jaws is clearly a sign of discontent and threatening behaviour.

All this shows us that the means by which wild

animals communicate with each other, by both smell and gesture, are more varied and difficult to interpret than we imagine.

Unlike most herbivorous animals, the rhinoceros (either African or Asian) moves its ears in a way which might seem puzzling. Ears flattened or laid back are a sign of contentment or great happiness. When pricked forward, they convey restlessness and imminent attack. Finally, a mysterious method of communication has been noted among African rhinos. It is not uncommon to find on ground over which a rhinoceros has passed, a strange semicircular imprint, for some two or three yards, apparently made with the tip of its hoof. So far, no zoologist has been able to explain its significance.

Do we know any more about the use of scent and the power of smell among big game?

Bears[7] are not content only to leave signs of their presence scattered around like so many notices advertising their private property. Every morning, they inspect them and top them up if they think the scent is not sufficiently strong. Here again, the poet who speaks of *odor di femina* is mistaken. In the great bear family (like the hippopotamus), the male alone has the right to warn other bears that they would be well advised to pass on.

Among tigers, lions and other wild cats, the sense of smell must be regarded as a means of expression, not only for information. Research is now leading to some surprising results.

We have long held emotional ideas about the unforeseeable accidents suffered by experienced animal trainers in menageries. Sometimes they are attacked and savaged by animals they have trusted. Is this due to a sudden desire for vengeance, to reassert their authority, or the onset of madness? These are possible explanations, but in most cases, the truth is more prosaic. A trainer, dressed in the usual costume with the same voice and the same professional gestures as always, enters the cage where his animals are waiting. Imagine that for some reason the floor is wet. The man slips. He is taken by surprise, but quickly regains his balance. Unfortunately, for a few seconds during which the incident lasted, the trainer was frightened. This simple reflex was enough to cause an immediate change in the acidity of his skin. If one of the animals, more sensitive than the others, notes this difference of scent, its instinct will be to attack. Because the man no longer has his familiar scent, the animal sees in him a stranger, a threat, a victim!

### The elephant's expressive trunk

Dr Wolfdicter Kühme, of the Max Planck Institute, patiently observed African elephants round the clock in the famous Taunus Zoological Park. Two-thirds of their territory is surrounded by the forest of Königstein, and its hills and valleys make it very like a natural reserve. By comparing their behaviour with that of wild elephants in Africa, he was able to interpret some of their gestures and attitudes which until then had not been understood.

"Thanks to him", writes Gerhard Gronefeld,[8] "I learnt elephant languages." Dr Hediger had already observed the obstinate behaviour of elephants in systematically destroying all man-made signs. However well signposts may be planted, they are invariably pulled up and boundary posts and fences are knocked down, no matter how solid the concrete.

How could the gentle elephant "tell" us any more clearly that he wants to remain free and to keep his boundless territory intact?

Elephants can express themselves easily with their flexible trunks and flapping ears. Their trunks end in two useful appendages as sensitive as fingers, and their ears are huge, representing in African elephants almost a sixth of the whole body surface.

So far our knowledge of elephants has been based mainly on the domesticated Indian elephant. It is so well trained that it can understand perfectly twenty different vocal orders and more than twice that number of gestures and simple contacts made by its mahout's feet or hands. But we are now beginning to understand the various elements of the "gesture language" of wild elephants, whatever their country of origin.

If an elephant is angry with a man it has no need to trumpet its anger aloud. If it merely wishes to show that it regards the man as an enemy, it will stop with head held high and ears spread wide, then walk rapidly towards him with its trunk hanging loosely. If it decides to attack, it swings its trunk in rapid movements in the same direction as its body. Some five yards away from its enemy, this threatening, swinging movement is at its greatest. But the elephant is not yet ready to use its trunk to inflict a heavy blow. Curiously enough, it must first lightly stroke one of the orifices of its temporal glands for a few seconds, before delivering an oblique, upward blow. The more aggressive the beast, the more violent is the blow and the more rigid the trunk.

Sometimes the attack takes quite a different form, even bordering on the semi-ridiculous. With lowered head and pendulous ears, its trunk half-forgotten, the elephant rushes on its enemy. Then suddenly, this huge mass, which seems too heavy to manoeuvre, turns round and quickly withdraws. Then it immediately raises its head, looks behind, pivots on its hind legs at an unexpected speed, at that very moment delivering a tremendous horizontal blow with its trunk at the unsuspecting victim, who had thought himself quite safe.

In this strange pantomime, the elephant's ears and trunk sometimes play an unexpected role. If the animal seems hesitant about its course of action, it bends its trunk gently and raises it to its forehead, seeming to scratch it, without conviction (mechanically, we might say), just as we scratch our heads when perplexed. If it decides to run away, it does so quite unashamedly. Laying its ears back against its skull, it breaks off the attack without further ado. But we have no idea of the reasons for its behaviour.

It is, however, much easier to recognise the cries, gestures and attitudes of a terrified elephant. Just show it a mouse, a jerboa, a dormouse, or even a hedgehog. The apparent illogicality of such a comic spectacle makes it plain that only the smallest advances are possible in our knowledge of animals, unless we admit from the start that even the most highly evolved mammal just does not see the world through our eyes.

## Notes

[1] Hontag, M.    *Psychologie du cheval*, Payot (Paris, 1954)
[2] Piéron, H.    *L'homme, rein que l'homme*, P.U.F. (Paris, 1967)

| [3] | Chavance, M.-J. | *Académie Vétérinaire*, XLII, Vigot Frères (1969) |
| [4] | Klopfer & Gambler | *Le comportement animal*, II, Crépin-Leblond (1968) |
| [5] | Sire, A. | *Les sociétés animales*, P.U.F. |
| [6] | Hediger, H. | *Nos amis exotiques*, Amiot-Dumont (Paris, 1954) |
| [7] | Hediger, H. | *Nos animaux sauvages*, Prisma (Paris, 1968) |
| [8] | Gronefeld, G. | *Understanding Animals*, Viking Press (1965) |

# CHAPTER II

# Almost human: communication among monkeys

Now we have reached the highest rung of the ladder.

Will monkeys, the most developed of all the higher mammals, cast new light on this vexed question of "animal language"? Among quadrumanes (which are neither true quadrupeds nor bipeds), we find new evolutionary components which clearly set them apart from other mammals – that is, hands and more highly developed brains. As we gradually ascend the simian scale, we find a noticeable hierarchy, even between members of the same species. Some monkeys remain on the plane of the lower mammals, while others, the primates, are capable of real intelligence. With the latter, the prehensile front limbs are more important than the hind limbs. In addition their more complex brains, equipped with a greater number of neurons, convolutions and connecting nerve fibres, increase their powers of association.

Among primates, the hand is no longer merely a means of locomotion, but is linked to the highest centres of intelligence. The sensori-motor nerves of the cortex provide a very delicate control of facial expressions, gestures of the hand and sharpness of perception.

With such a wealth of cerebral equipment, will monkeys ever be able to articulate and gesticulate like human beings? The answer is "yes", but only "in appearance".

Monkeys can, in fact, cry, moan, shout and murmur because they use their vocal chords and lungs as we do. Their mime is often undoubtedly appealing and sometimes most striking as the muscles and nerves of their faces, jaws, lips and tongues work in exactly the same way as ours. Nevertheless, a great gap still divides us.

The chimpanzee, considered to be the most skilful and intelligent of all the great apes, can even write and paint (or rather, scribble and daub) but he does not know that he is "writing", any more than he knows that he is "painting". He will use, in his wild state, tools which we ourselves might use (such as dry twigs which he gathers near ant nests and uses to unearth the termites), but he could never begin to invent a tool.

Enough of this rather tedious description of the brain's anatomy and functions. Only expert biologists can explain the reasons why a monkey is incapable of the notion of real symbols. It suffices to say that even in the most advanced great apes there does not exist, any more than there exists in the common macaque, a "device" corresponding to the cortical framework which provides man with the ability to express himself in human language. It should be stressed that, from the marmoset to the gorilla, not a single monkey has ever attempted to imitate our spoken language, although we all know how easily they can imitate our gestures.

## Monkey tricks and antics in the wilds

A wild monkey is always on the move. Everything in his surroundings suggests threats, fears of the unknown. Everything both attracts him and frightens him off, but however great his curiosity, he rarely stays long enough to satisfy it, because something else claims his attention. Hence the reason why the behaviour of these crying, whistling, shouting, gesticulating creatures is so hard to monitor. For a long time it was possible only to observe that wild monkeys would flee at the slightest noise and to note their warning cries and the way their young ones called for help. It is impossible to observe dog-faced and other baboons because of their noisy jabbering.

Also, in reality much more placid and sober, the great primates (such as gorillas and orang-outangs) were long regarded as dangerous creatures, to be approached only when armed with a gun. For that reason we have so many stories, legends and eye-witness accounts, all pure fantasy, but which, sad to say, are far more numerous than serious observations. Fortunately, over the last few years, thanks to the cine-camera and tape recordings, authentic documentation has become possible. We know now, for example, that not all monkeys are totally vegetarian. South African baboons frequently attack sheep and antelope. After immobilising their victim they rip it open with their teeth, tear it to pieces and fight over the body even before it is dead. This sort of behaviour (unknown until now) belongs to social monkeys living in herds or family groups. We now have a better idea of their group structure, their display of sexual or parental instincts, and the real explanation of that traditional habit of so-called "grooming". This is not simply an ordinary peaceful method of cleaning, but it has its own special role in the monkey world.

It now remains for us to give a precise meaning to all the lip movements, solemn greetings, displays and disconcerting postures, which, together with vocal sounds and gestures, make up "ape language".

How can we understand the significance of the medley of noisy cries made by fifty or so howler-monkeys leaping from branch to branch, apparently abusing each other in piercing yells until the one which shouts the loudest takes precedence? Always living in a state of sexual anarchy, these crazy acro-bats only seem to recover some sort of calm and orderly family life when once in a while they risk leaving their tree-world for the ground.

Baboons often assemble in groups of a hundred-strong, divided up into families, each a more or less permanent establishment of a male, five or six females and their many offspring of different ages. Taking refuge in narrow valleys and in the rocky hollows of ravines, they spend most of their time in conscientious "grooming" and passionate embraces. Not all of these have results, but in this way certain ardent females, who are in great demand, attain high rank in the group.

It is among baboons especially that one finds a great range of extraordinary sexual displays, whose exact meaning escapes us. For instance, their manner of saying "Good morning" is to stop in their tracks and display their buttocks. No less odd is the strange behaviour of the male head of a squirrel-monkey family towards a sexually inactive male: he will spread his legs and exhibit his penis in a symbolic gesture, a demonstration which goes no further. Thus challenged, his fellow-monkey squats down, lowers his head to protect himself from blows, and keeps quiet.[1]

One is tempted to compare this admittedly rare animal gesture with a similar human gesture. In certain North African countries one of the most humiliating and provocative gestures possible is for a man to raise his arms in two or three jerky movements towards his body.

To display the posterior, which is commonplace among monkeys, is on the other hand, a sign of self-abasement and allegiance. Among females this gesture does not necessarily signify a willingness to couple, for it is also the expression of her humility in the male's presence and in certain cases, the admission of the superiority of another female. Among young monkeys, it is always a sign of submission and weakness.

Wild macaques or barbary apes, as numerous as baboons, have also been studied by experts with extremely interesting results. Japanese macaques, for example, will not tolerate on any account the sexual promiscuity practised by baboons. Junichiro Itani reports that the great chiefs of these Far Eastern monkey tribes are so jealous of their rights that they will scarcely permit even a temporary delegation of authority to those chosen to keep order in their absence.

This superiority of dominant monkeys is certainly a question of power and aggression, but it is also a matter of teeth! Never has the expression "bared his teeth" been more applicable than when a monkey pulls that classic face which serves to command instant respect.

This method of intimidation is found among all dog-faced baboons whose impudence is only equalled by their instinctive aggression. Vervets prefer to live in rocky regions and occasionally in the rocks themselves. Perhaps they tend to avoid man and make no attempt to earn our affection because we rarely treat them with the same amused indulgence shown to other monkeys. Certainly mandrills, drills and hamadryads (Arabian baboons) have cruel, obstinate, insolent expressions – enough to put off the most kindly disposed person, particularly if one knows their furious blind rages which can break out in a few seconds. Moreover, mandrills possess certain peculiar anatomical features which add to their disquieting appearance. For one thing, their faces are outrageously coloured blue and red like a war mask or carnival disguise.

What patience and self-control must therefore have been exercised by Professor S. Washburn and Dr I. de Vore, two animal psychologists from the University of California, to mix with a troop of some hundred dog-faced baboons in the Amboseli savannah (at the foot of Africa's Mount Kilimanjaro). So successful were they that they eventually managed to interpret some of the gestures and grimaces of these irritable monkeys and at the same time discover some of their habits.

Like most monkeys in their wild state, the social life of dog-faced baboons is founded on gregarious instincts. Among these apes, leaving aside all "morals" and despite their promiscuous sex-life, there exists at the very heart of the group a perfectly organised communal life. An old, experienced male reigns peacefully over his fellow apes, each of whom has his proper place and status in the community. Each one knows his duties and his rights.

To convince themselves of this fact, Washburn and de Vore followed day by day the individual behaviour of these apes. Tolerated without hostility, the two zoologists were soon able to note certain specific characteristics simply by distributing some titbits. Each time they threw a piece of food between two monkeys who appeared to be about the same age, build and strength it was the senior of the two in rank who, without any argument claimed it. A simple authoritative glance would be enough for the other to withdraw without further ado. But what if the latter were to show signs of insubordination? A low growl would send him hurrying back to the ranks. If he still persisted the senior monkey would crouch down like a watch-dog ready to pounce, his fur on end, and lips drawn back disclosing his teeth. If defiance still continued, the injured party would then strike the ground sharply with his right hand as though telling the other to mind his manners ("That's enough now! Let's have no more of your insolence"), or else he would strike the ground with a stone, but would never use it as a weapon or missile. Indeed, as the writers make clear, bloody fights between these dog-faced apes are distinctly rare.

This would suggest that these monkeys possess a "language" consisting of accepted gestures, whose nuances they all understand so that anger can never degenerate into single combat.

But what if more violent emotions are aroused? What if there is a threat to the whole community, if the presence of a wild beast is signalled by some alarm call? All disputes die down at once. Each one quickly takes up his position, for either defence or attack. The adults surround their leader; the females and the youngest retire to the rear, while the males form a front line with such yells and gesticulations that, in nine cases out of ten, the enemy does not press his attack.

Once the danger is over, there follows a cacophony of cries: groans and smacking of lips and tongues, accompanied by wild leaps into the air, frowns and grimaces – strangely similar to the emotions displayed by a crowd of jabbering so-called civilised beings commenting on an accident or a riot.

What exactly, then, are these apes saying? Obviously we do not know. However, Vitus B. Dröscher, who has also witnessed such events, claims that these vocal emissions are not devoid of rhyme or reason and he reports the following as proof. One day at the beginning of a violent storm, a tape recording was made of one of these "discussions" at the Bronx Zoo in New York. A few days later, when the sky was blue and the sun was shining, the tape was played back. The same monkeys questioned themselves in surprise for a moment, then dashed hurriedly away into every available corner to shelter from the rain which according to their own recorded "words", would soon pour down.

But perhaps this is to put to much faith in the possibilities of "language" when associative memory, or perhaps nothing more than fear caused by a strange noise, would explain their behaviour.

## A woman among the chimpanzees

The observations made by a young English student, who is deeply interested in anthropoids, are very different. In 1960, Miss Jane Goodall (now Mrs van Lawick-Goodall) had the good fortune to be accepted by a large troop of wild chimpanzees, living on the shores of Lake Tanganyika.[2]

Every study of animal psychology quotes her strange adventure. For months, Miss Goodall observed the trials and tribulations of a group of fifteen of these anthropoids as she followed them at a distance of some two hundred yards. She had managed to reduce this distance by about half without provoking any reactions, until one day, in the forest, she suddenly found herself surrounded by six large adult males, all gesticulating and furiously excited. When one of them hooked himself on to a branch above her head and began to swing to and fro, uttering fearsome cries and making terrifying grimaces and gestures, Miss Goodall, half-dead with fear, threw herself flat on the ground, unable to move. This instinctive gesture proved to be her salvation: without realising it she had clearly expressed her allegiance in "chimpanzee language". Reassured by her attitude, the frightening anthropoids calmed down and began quietly to nibble fruit, without paying any further attention to her.

The young zoologist courageously returned the following day. The chimpanzees continued to accept her presence without any signs of suspicion. In this way, she was able to familiarise herself with these primates, day after day for four years, to such an extent that she dared shake the hand held out by one of them.

Just as in fairy tales, chance had stepped in to present her with an outstandingly important break-through – the unhoped-for key to a sympathy between man and beast. Was this spontaneously outstretched hand offered to a creature of another species the benevolent gesture of a superior to an inferior, or was it a humble gesture begging for something to eat? Both explanations are possible.

Whatever its meaning, this joining of hands assumed the ritual value of a pact, as Mrs van Lawick-Goodall was able to prove after months of experiments and observations. In fact, after a time, by copying the mime and gestures of her mysterious friends, she was not only tolerated, but admitted into the troop. Little by little, she was even allowed to play with the baby chimpanzees, provided she respected a rigid code of behaviour. Before touching the babies, she had to give a sign of friendship. This consisted of slowly stretching out her arms with the palms of her hands turned upwards, while she stared calmly into the eyes of the mother, awaiting the latter's look of assent.

Thus everything began as a result of a simple gesture. It was such gestures – peaceful symbols, obeying some secret law – which later inspired zoologists elsewhere to make a first attempt to estab-lish communication between human beings and chimpanzees by means of deaf and dumb language. "Establish confidence first and communication follows." No doubt we need seek no further for the origins of the traditional "handshake", which today has become as mechanical as it is carelessly bestowed and which still continues to be exchanged between human beings everywhere after thousands of years.

We will see later how far this gesture of the proffered hand was to prove the key to future discoveries. It was to allow other research workers to continue in greater detail the work begun by Mrs van Lawick-Goodall on the behaviour of chimpanzees in their native environment, and to confirm her conclusions.

At about the same time, a Dutch scientist, Dr Adrian Kortlandt, also tried to decipher the gesticulatory code of chimpanzees during his many visits to Guinea and the forests of the Congo.[3] First he drew attention to the fact that the outstretched hand had several meanings. For example, when two primates meet, if one turns the palm of its hand downward, this signifies obedience. If it places it hand in that of the other, this implies agreement and equality. The use of the index finger by itself is a silent warning of imminent danger. Arms raised in the air means, "Come no further", or if crossed, "The way is clear, but be careful!"

Many other gestures have been noted, but their meaning still eludes us. Together with facial expressions, all these gestures and cries apparently constitute inborn methods of exchanging indispensable information or of communicating vital needs.

This, at least, is true of chimpanzees and for rather special reasons. Chimpanzees, which are considered to be the most intelligent of all simians, remain under maternal influence for a very long time. For seven years (sometimes even longer) a chimpanzee remains a child both in the eyes of his own family and to those about him. How, during this long period of family instruction, can he fail to learn – even before the sexual instinct is felt – all the vocal sounds, postures and gestures which will eventually allow him to find food and shelter for himself?

## The other great apes

There remain two other great apes – the gorilla and the orang-outang, both of which are as little-known as they are disliked. These huge anthropoids have not the same excitable disposition as chimpanzees. On the contrary, their bearing resembles that of wise old men plunged in profound thought. Their calm, unblinking gaze, their size, weight and strength have unjustly earned them the reputation of being saturnine brutes.

I personally have never had the chance to observe or catch unawares a single wild gorilla, but I have nursed one which had fallen seriously ill after being captured. Later, it was to become the strongest and heaviest (420 lb.) of all known anthropoids in captivity. Despite the often disagreeable and sometimes painful treatment which it suffered at my hands, it never showed me the slightest signs of hostility.

It is claimed that in the wild state, these impressive-looking primates are astonishingly gentle, extraordinarily brave, and will only attack in self-defence. They never attempt to flee in the presence of man and are only afraid of dogs. Some animal psychologists have been struck by a form of mime which is more eloquent and significant among gorillas than among any other mammals: a habit of staring eye to eye. More than all other insulting behaviour found among the great apes, this is the supreme expression of a challenge.

This is perhaps not much to go on if we wish to discuss their habits or embark on a study of their methods of communication and the means by which they express themselves. But imagine the dangers, or at least the difficulties, of carrying out a continuous, day-to-day study of several families of gorillas, who are moving around completely independently, in the depths of the forest.

## A man among the gorillas

Few research workers have actually risked such an adventure. To our knowledge, only one work has been devoted entirely to the study of gorillas in the wild state.[4] George B. Schaller, a young American zoologist, in fact spent a whole year among gorillas in the mountainous forests of the Congo. By dint of patience, and because he could respect the total freedom of the subjects under observation, he succeeded in being admitted into the intimate life of these giants, who could have flayed him alive as easily as we peel the shell off a shrimp. For months on end he lived in the "musty, somewhat sweet odour of gorilla" and experienced the same feeling of fear every time he heard "the explosive, half-screaming" roar which petrifies the boldest of explorers. Sharing the life of groups of eight to twelve gorillas, he was able to film them and record their slightest sounds. His conclusions, therefore, are of the greatest interest as far as their "language" is concerned.

"The apes", he writes, "seem to lack the tendency to vocalise for the sake of vocalising, a trait which is so important in man. No infant gorilla ever babbled like a human baby. The gorillas had no interest in imitating sounds, in practising with various combinations of sounds. . . . This is not to say that gorillas lack a way of communicating with each other and that their method is not perfectly adequate for their simple mode of life. But the gorilla's ability to impart information to a neighbour is confined entirely to the situation at hand; there is no way to convey something that happened yesterday. On the whole their signalling system is no more complex than that used by dogs and many other mammals. Gorillas co-ordinate their behaviour within the group primarily by employing certain gestures and postures. For instance, a dominant male who walks away from a rest area without hesitation imparts not only the information that he is leaving but also the direction he intends to take. In order to be groomed, a gorilla merely presents a certain part of its body to another animal. Each gorilla simply keeps its eyes on the rest of the group most of the time and does what the others are doing."

What struck George Schaller was the near-dumbness of these huge apes whose mime is so expressive. The voice, which is of such importance in our own society, seems to be of secondary importance to them. So far, scarcely twenty distinct vocalisations have been noted and recorded, and only eight of these seem to be in current usage. A low grumble signifies satisfaction; abrupt grunts are used as a guide when a group, scattered among the vegetation, are worried that they have wandered too far from one another. Lastly, there are the deep-throated barks which gorillas utter when irritated and a sort of screaming roar they make when angry.

It seems unlikely that all these sounds are made voluntarily to establish true communication. The members of a group have learnt that certain sounds correspond exactly to certain situations: these precise vocalisations have, for them, assumed the value of signals.

"Although the number of basic vocalisations emitted by gorillas is fairly small", George Schaller emphasises, "there is considerable variation in the pitch, intensity and pattern of each sound. These variations greatly broaden the scope of the vocal repertoire, for the animals respond selectively to the sounds they hear. Their reaction depends not only on the sound but also on the condition under which it is given and the member of the group who gives it. For example, no member would confuse the deep full grunt of a male for that of a female."

Moreover, Schaller noticed that a certain female habitually cried out each time he approached her group. Yet the other gorillas paid scant attention to her cries, even when she was out of sight. This clearly showed that they recognised the voice of the female concerned and her stupid habit of shouting "Help" for no valid reason.

One sound may have two equally valid meanings depending on circumstances. The hoarse growls of the chief, putting a stop to a female quarrel, have a certain significance. But if this male makes the same grunting sounds when there is no such incident, all the members of his group look at him and spontaneously turn their heads in the direction of his gaze.

In spite of their human appearance, the way gorillas move gives them the look of subnormal men wrapped in fur coats, their legs too short, who wake up, stretch, yawn, eat, rest flat on their backs with arms folded under their heads, while their young ones fly into a temper when unable to have their own way. All this is far removed from the complicated habits of the chimpanzees, with their constant state of excitement and their flashes of intelligence.

Surprising as it may seem, there is even more

difficulty in studying a gorilla in captivity. The subject must be very young, for after the age of five, the gorilla's development is retrogressive. The little gorilla prodigy, whose rate of mental development is astonishing compared with that of a human child, then returns quickly to being nothing but a stupid brute. Anatomically speaking, profound differences separate the development of the two. The cranial sutures (i.e. the serrated articulations of the skull) are completely fused between the ages of five and six in gorillas, while in man this only happens at around the age of thirty. Thereafter no increase in brain size is possible.

Nevertheless, we have noted several peculiarities in the behaviour of these giants which indicates their low level of intelligence. For example, they will not look for fruit trees beyond their immediate environment. Laboratory tests demonstrating such a fact explain the inability of gorillas to use images. Gorillas are also indifferent to motionless objects in their environment, while, as W. Welker proved in 1961, they show relatively little interest in new objects and soon become bored with them.

One can imagine the difficulties encountered when attempting to study the subject of communication between captive gorillas, if only because of the rarity of these great apes. Such studies, therefore, have been almost exclusively limited to relations between gorilla and man, and man and gorilla. Even so, scientific observations on the subject are exceptional. Even in the wild state, the gorilla is shy and makes gestures as rarely as sounds. Only his facial expressions give us any idea of his feelings (or more precisely of his emotional state), which he can never conceal.

To some up, with the exception of Miss Cunningham's well-known John Daniel and Yerkes's celebrated Congo, gorillas are virtually untrainable. However, they are probably better able to interpret what their keepers say to them than we can interpret their vocalisations or their even more mysterious gestures, which are impossible to consider without running the inevitable risk of giving them a human meaning.

## The Asiatic orang-outang

More placid by nature, more contemplative and even less vocal than the gorilla, the Asiatic orang-outang from the isles of Sundra and Sumatra differs radically from other great apes. The principal emotional characteristics noted by different observers (including Robert and Ada Yerkes)[5] are, in more or less human terms, their sense of fear, self-effacement, unpretentiousness, shame, terror and even modesty. From a state of apparent passivity, an orang-outang will suddenly fall into a rage or become panic-stricken. Impatience, fury and malice lead to emotional scenes during which it beats its head on the ground or stamps with rage, reminiscent of the behaviour of an angry human child. This behaviour, together with mimicry and gestures, has all the eloquence of a language. In fact, an orang-outang indulging in this sort of pantomime has been seen to interrupt its performance, then resume, according to the effect it is producing on its audience, whether human or fellow ape.

Regarding the production of sounds, animal psychologists and anatomists are inclined to see a direct link between the vocalisation of these great primates and the laringeal cavities of their throats. These unusual cavities have been carefully dissected (by Camper in particular), but what part they play in the oral emissions of these great Asiatic apes is still a subject of dispute.

The only valid contribution to the possibilities of communication among orang-outangs seems to be that which was made by Furness. According to this writer (famed for his patience in making an orang-outang, after 6 months of daily training, pronounce two words), orang-outangs are perfectly capable of acquiring a surprising degree of "understanding" of the human language. But are they also capable of learning to use this language? The first ape to pronounce the word "papa" was certainly exceptional. When this famous orang-outang was asked, "Where is papa?" She would point her finger at her master, or touch him on the shoulder. Then, one summer's day, when Furness had taken his pupil in his arms and was attempting to give her a bath in the swimming-pool, the orang-outang, feeling the water rising up her legs, desperately hung on to her master's neck and repeatedly called out: "Papa, papa, papa."

It was tempting to make the link between this pathetic cry and its real significance. Unfortunately, we still know too little of the emotional behaviour of orang-outangs (who remind us, in so many other ways, of Chinese pygmies) to draw any solid conclusions from this true story. All we know definitely about Furness's affectionate, thoughtful and good-natured great ape, is that it died a short while later after having also learnt to say the word "cup" when it wanted a drink. We can only regret that such en-

couraging and promising experiments have had no
follow-up.

## Notes

1 Morris, R. and    *Men and Apes*, Hutchinson (London, 1966)
   Morris, D.
2 Goodall, J.    "My life among wild chimpanzees", *Nat.
   Geog.* (August, 1963) pp. 272–308
3 Kortlandt, A.    "Des bases physiologiques de la culture",
   *Current Anthropology* (1965)
4 Schaller, G. B.    *Year of the Gorilla: An Exploration*, Penguin (London, 1967)
5 Yerkes, R. M. and    "The Orang-Outang", *The Great Apes: a
   Yerkes, A. W.    study of anthropoid life*, Yale Univ. Press
   (New Haven, Connecticut, 1929)

# CHAPTER 12

# Almost human: communication between monkeys and men

No one questions the intelligence of monkeys. At an early age, they develop faster in a human environment than in their natural surroundings. However, the limit of their mental capacity is soon reached. At a maximum age of three years, the anthropoid ape reaches a full-stop. It can develop no further.

The numerous works devoted to this subject, attempting to establish a parallel with the awakening and blossoming of intelligence in a human child, and trying to recognise the possibilities of communication between the primates and ourselves, have all led to identical conclusions.

N. Kohts[1] describes the conclusions drawn from his daily observations of the parallel development of a young chimpanzee (Yony), and his own son (Roody) who was exactly the same age. The study was made between the ages of eighteen months and four years, and was not concerned with training or teaching, but purely with objective research. His detailed literary description (illustrated by drawings and photographs) would allow psychologists and zoologists all over the world to follow, in minutest detail, the behaviour of the two subjects.

It was invaluable to know how the animal and child were able to communicate with each other. Mrs Laduigina-Kohts noticed that the two babies spontaneously developed a conventional language of gestures, curiously similar in both. An outstretched hand expressed demand; to turn away the head meant refusal of food; to raise the hand to the mouth expressed thirst; to tug at the clothes of someone nearby was to demand attention; and so on.

But which of the two, the monkey or the baby, copied the other?

Very early on, Yony, the young chimpanzee, began to utter all the noises which seem to make up the repertoire of its species, among which one could distinguish twenty-five separate sounds. These can be summed up in a few words. Excitement is expressed by a series of paired modulations rather like the barking of a dog; anger is close to this, phonetically, but ends in a dry bark; distress by a single, prolonged moan, a sort of "Oooouh! . . . Oooouh!" pronounced with very distinct inflexions. For example, if the "ooh!" sound is long drawn out, this expresses lassitude and if the "uh" is emphasised it signifies displeasure. Extreme unhappiness is expressed by a moan which Kohts describes as resembling both a cry of pain and the sound made when one strikes a cracked glass. Finally, fright is conveyed by a very short muffled "ouf", while satisfaction or relief is expressed by successive gurgling noises "Khriouou . . . Khriouou . . . !"

Mention could also be made of the feeling of

# Galaxy Of Awards Prize Levels

| Level | | Prize |
|-------|---|-------|
| Level | A | $15.00 |
| Level | B | $30.00 |
| Level | C | $50.00 |
| Level | D | $70.00 |
| Level | E | $100.00 |
| Level | F | $150.00 |
| Level | G | $250.00 |
| Level | H | $400.00 |
| Level | I | $500.00 |
| Level | J | $750.00 |

vexation (which in the author's opinion is demonstrated by hoarse accelerated breathing) and the sensation of an agreeable taste (expressed by soft moaning sounds).

What are we to make of such observations? On the whole we can confidently state that the chimpanzee imitated the tone of the sounds made by the baby without actually articulating them. It was simply accentuating sounds in which it seemed to recognise some affinity with its own.

As for the child, such sounds (which serve to express emotional states) already formed part of little Roody's vocabulary at the age of seven months. At eight months he uttered his first complete word.

Where conditioned reflexes are concerned, the little chimpanzee obeyed the following twenty-four verbal orders: "Come here", "Go away", "Sit down", "Lie down", "Turn a somersault", "Play ball", "Get up", "Climb on the cage", "Give me your hand", "Play", "Back to your place", (he would then sit on a chair pointed out to him only once), "That's forbidden", "Leave it" (he would stop touching the forbidden object), "Give me the ball", "Off you go", "Let's go for a walk" (at which he would put out his hand), "Give me your blanket" (he ran to get the blanket), "Give it to me" (he would hand his master the first object he came to), "It's hot" (he would then touch the object indicated with great caution), "Get in your cage" (he would obey), "I'll be back in a moment" (he gurgled with pleasure), "I'm going out" (he would cry but never shed a tear), "The wolf is coming" (he would be frightened and seek safety in his cage), "Eat!" (he would do so), etc.

The experiment obviously could not fail to impress other research workers.

At about the same time, in America, the Kelloggs tried an identical experiment. Having brought up a young female chimpanzee named Gua with their own child Donald, in the space of sixteen months they succeeded in teaching this she-monkey fifty-eight English words to which she reacted correctly (even though several were more or less independent of tone of voice or gesture). This time it was truly a question of an acquired "language", a code which had nothing in common with the instinctive language of monkeys, but was the result of training.

A little later, two more scientists, a Mr and Mrs Hayes, also undertook a similar study.[2] The experiment with a young female chimpanzee was to last three years. During that time, Vicki, the little monkey, performed much better in tests than human children of her own age, but then all progress stopped. Just like the Kelloggs' young Gua, Vicki also learnt to obey orders like "Shut the door", "Give me my stick", and "Sit down". But try as they might, over months and even years, the Hayes did not succeed in teaching her to say words.

During seven years of family life, Vicki, who was without doubt one of the sweetest-natured and most diligent ape-pupils ever studied, managed to say only two words – "cup" (which she associated with thirst and which she repeated continuously until she was given something to drink), and "Mama" (to which she certainly attributed no meaning, but which she pronounced as a vocal display of happiness when she was snuggled down in the arms of her mistress).

Time passed. It had to be admitted that anthropoids (and *a fortiori* the inferior monkeys) possess only a very limited inborn vocabulary with which to communicate orally with humans. Perhaps they would prove better endowed with a language of gesture?

## Laughter is unique to man

Monkeys are essentially gesticulating creatures, forever moving and grimacing. Mimicry plays a major role in their lives. The fact that Mrs Kohts was able to trace ninety-five wrinkles on the face of her sleeping chimpanzee shows just how flexible simian features are.

This mimicry, however, should not be considered equivalent to human mimicry. For example, the grimace which could be interpreted as an expression of mirth (mouth wide open, disclosing the teeth, with lips drawn up at the corners), corresponds more to fear or anger than to joy. The kiss, which, in babies of two or more years old, is a spontaneous expression of affection, is a completely artificial act among monkeys and is very rarely seen.

Nonetheless, the Kelloggs claimed that Gua could smile and laugh. It is true that, from the age of eight months, their little chimpanzee, when tickled under the armpits, "gurgled" with rapid little choking gasps and when she was about twelve months old, this phenomenon was accompanied by a sort of stifled laugh when someone pretended to be about to tickle her. But is this sufficient to justify speaking of animal "laughter"?

Buytendijk clearly accepts the possibility of laughter among these higher animals, including anthropoids. The chimpanzee can at least mimic a laugh which could be interpreted as an indication of joy or playfulness in the strictest meaning of the

187

words. An animal's laugh is related to play. The effect of tickling on a young chimpanzee seems to be a confusion of the agreeable and the disagreeable. This produces a spasmodic tension which is relieved by a rictus. When, later, the little monkey is threatened with tickling, it anticipates the situation, hence the deceptive mimic laughter. This, however, is different from true human laughter, which is an expression of conscious inner joy.

Some lovely photographs of apes have provided us with evidence of this mimicry and its apparent meaning which nevertheless disturbs us. In fact, simian laughter could well be confined to chimpanzees, for similar signs are never observed on the grave countenances of gorillas or of orang-outangs.

We have often noticed that, as well as the most intelligent primates, dogs also mimic laughter. But here a profound difference must be noted. While a dog, in contact with a human being it loves, remains an eternal child – happy and playful until it reaches a relatively advanced age – the higher ape, once fully grown, no longer laughs and scarcely ever plays. The most mobile and frolicsome chimpanzee is therefore, in the words of Buytendijk, like "a tragic child, who after a period of escape is condemned to return once more to its proper animal nature. . . ."

## Hands and feet

From the point of view of gesture, quadrumanes are obviously better equipped than other mammals. They can use the extremities of all four limbs quite freely. But can they use them as a means of social communication?

Professional interest cannot fail to be aroused by Yerkes's famous and much-quoted observation. One day, Moos, Yerkes's favourite chimpanzee, fell ill, or at least refused all food except purees and liquids. Thinking that there could be an infection of either gums or teeth, an assistant entered the cage to examine the patient's mouth. Moos made no objection and no abnormality could be discovered. The assistant was about to leave when Moos stopped him by pulling at his coat. Moos then raised his upper lip with one hand and with the other pointed to a very precise spot on his gum. On closer inspection there was found to be a slight swelling, caused by a growing tooth, which made chewing painful. "One can understand", says Yerkes, "that our vet felt some resentment at being helped in his diagnosis by the patient himself."

Joking apart, if we look more closely at this rather surprising gesture made by a simple animal, albeit one of the higher primates, we can see that this was, after all, just the most elementary way of showing the association of feelings or actions. What does it matter whether or not Moos indicated the exact seat of his pain? This gesture made by an intelligent animal is clear enough to us: in exceptional circumstances, when confronted by something unusual and worrying or something which causes pain, higher mammals, such as anthropoids, are perfectly capable of finding a means of helping themselves (a tool, even). In such a situation the chimpanzee was behaving exactly like an irritated dog when it barks in front of a closed door. Moos may not have appealed to the man's understanding, in the abstract meaning of the word, but, using the vet as a "tool", he "associated" it with his pain, and that is why his gesture can, to a certain extent, be considered as a form of language.

How can we explain the following incident, reliably witnessed by Hilaire Cuny[3], which took place at the Vincennes zoo? It concerns Pépée, a large female chimpanzee, who was popular with the crowds, and all the more spoilt by them because she usually walked about on her hind legs. She frolicked around in semi-liberty, only separated from the crowd by an impassable ditch. A notice indicated that it was forbidden to feed the chimpanzees, but regular visitors to the zoo paid little attention to this warning. Pépée definitely understood what this meant, for, standing up on her hind legs, she would look at the keeper with a comical expression and then, as soon as the man had turned his back, she would immediately indulge in an unmistakable pantomime for the benefit of her friends. Her glance would pass from the keeper to the amused crowd and from the crowd back to the keeper, accompanied by gestures implying "Come on, come on! Give me something to eat quickly while he's not looking!"

Anthropomorphism perhaps, but Pépée's case was surely a display of intelligent communication between species.

## An experiment with deaf and dumb language

The various close studies made of monkeys throughout the world have inevitably led animal psychologists to regard the higher order of ape as a sort of deaf mute with whom it might be reasonably hoped one day to communicate by gesture. Therefore in

1967, two young teachers from the University of Nevada, R. Allen and Beatrice Gardner, decided to attempt a new type of experiment.[4] They knew that the chimpanzee's vocabulary was relatively poor. The twenty-five sounds or cries made by primates, as noted by Kohts, and representing a call for help, summons to play and to assemble, and an order to flee, were not an easy code to use. The apes in fact employ it only in moments of extreme agitation. Why not therefore attempt to teach a young "student" chimpanzee "the language of gesture"?

For this purpose the Gardners chose a young female, captured in the wilds, which had never yet come into contact with humans.

When she arrived at the laboratory, Washoe must have been about one year old. Allowed to eat when she liked and to take long rests, she was completely undisturbed by unnecessary words. Assistants and keepers were instructed to say as little as possible and to use only the sort of words one would say to a baby when feeding, calming or dressing it. Thus everyone avoided causing oral excitement, for monkeys, with their hypersensitive ears, are particularly susceptible to the slightest sound. Meanwhile, the Gardners learnt the deaf and dumb language used in the USA and made every effort to communicate with each other and their assistants in sign language only.

This atmosphere contributed greatly to the experiment's success. Washoe quickly and voluntarily accustomed herself to this silent language and with her natural instincts for imitation, spontaneously copied gestures of the people around her.

All they now had to do was impress on her mind this new form of expression, which was to be a method of inter-species communication. But since signs conveyed by fingers and hands have more to do with the language of analogy than with pure abstractions, American sign language could not be used with an animal in the same way as it is used daily by human beings. The Gardners and their team therefore perfected a slightly different code consisting of a series of word signs. As one can imagine, training the chimpanzee was not to prove easy.

First, it was necessary to train Washoe to discipline her natural gestures and adapt them to the gestures of her instructors. To achieve this the little chimpanzee was familiarised with everyday human life. She ate at the same table as her masters and took a bath each day. As she played with a doll, she soon came to imitate everything Mrs Gardner did. She would turn on the taps and turn them off when the water reached its usual level. She would then plunge the doll in the water and dry it, as seriously as you please, before putting it to bed, just as Mrs Gardner did with her.

By similar imitative methods (not to be confused with an awareness of her own "self"), she was taught to brush her teeth. Being inquisitive and extremely perspicacious, like all monkeys, she associated her mistress's sign (index finger imitating a toothbrush in use) with a real action. The action of brushing teeth in itself did not please Washoe, but in the end she complied with sufficient good grace to earn her mistress's (silent) congratulations and caresses.

The education of the little chimpanzee had been going on for about ten months when a flash of intelligent action lit up the darkness of her animal life. Going with her mistress into the bathroom of a friend's house, her attention was attracted by several toothbrushes. Recognising their appearance if not the symbol, she raised her index finger to her mouth and mimed the daily act of cleaning her teeth. Here for the first time was the undeniable wish to communicate a "thought" by means of gesture to a living creature of a different species.

From then on, there was a succession of experiments. Some were successful and some were not. Washoe indulged in a sort of gesticulatory "prattling", as her sign language vocabulary was gradually enriched. She babbled as she tried to link sign-words together. The Gardners, more anxious than ever that her gestures should not be ill-formed (either hesitant or incomplete), redoubled their patience. From now on, by making use of their little pupil's own instinctive gestures, they could make coded signs more easily.

The Kellogg experiments and the observations made by Mrs van Lawick-Goodall in Tanganyika have already shown that a primate's outstretched hand means: "Give! Give me!" It is the same gesture as that used by humans; it is the natural gesture of someone begging. Therefore it was easy to teach Washoe to perform this same gesture and then to complicate it a little more by expressing another idea: "Come!" (to ask for things which were out of reach, or to make them come to her). This was the same sign of the outstretched hand, but with the hand immediately flexed from the wrist towards the forearm – the same action as used by the deaf and dumb.

Good use was also made of Washoe's uninhibited pleasure when tickled. As soon as someone stopped tickling her, she seized the person's hands, and placed them round her neck or in the place where she had been tickled, as much as to say "more, more". This natural reaction made it possible to teach her to express the new idea of "more", "again" or "still more". For this, it was enough to join the fingertips of one hand, palm upwards, rather as Arabs will do to show their appreciation of what they are saying, or to express repletion or complete satisfaction.

The instructor repeated the gesture tirelessly until Washoe succeeded in performing it perfectly. Then, and only then, did she get her reward in the form of a titbit or the tickle she so enjoyed. In this way, use was gradually made of all her natural gestures and grimaces.

For example, Washoe loved to be thrown into the laundry basket. To encourage her masters to continue the game, she would give them an imploring look. From now onwards they made a point of replying to her using the sign "more". Thus she soon realised that it was enough to bring together the tips of her fingers for the game or caresses to continue. A month later she was using the sign "more" when she wanted more food or drink, or to continue with any other activity which she thought was coming to an end, like staying in the garden or watching television.

She was then taught another sign meaning "open". This consisted of joining the hands palm to palm, and then opening them from the wrist as one opens a book. Just as quickly, Washoe learnt to use this gesture whenever she wanted something opened – doors to rooms, the refrigerator, cupboards or boxes. Even simpler was the acquisition of the sign "key" – the tip of the index finger of the right hand turning several times (like a key) in the open palm of the left hand.

A little later Washoe managed to learn the sign

194

"flower": the tip of the index finger touching the nostril. She was then taught to distinguish between a dog and a cat and consequently to "speak" of them, but the presence of either of these animals excited her too much to use them as subjects for demonstration. Washoe could easily recognise either animal in a photograph or drawing. This image was associated with the following mime. Several quick movements of her tongue towards the inside of her thigh were to be interpreted as "dog". Washoe understood this sign so well that she voluntarily performed the action the first time she happened to hear a dog bark. The sign for "cat" was to pinch the cheek between forefinger and thumb.

In short, after two years, the gesture-vocabulary of the little chimpanzee had reached the not inconsiderable figure of thirty-four word-signs. Would it be true to say that she had succeeded in making phrases with these word-signs? The idea seems unlikely. However, Washoe began of her own accord to associate several gestures like "Give me", "Key" and "Open" to give a complete phrase meaning "Give me the key to open the door". Afterwards she learnt, as a luxury, to add a gesture which meant "Please".

Obviously one must not consider this addition as a useless polite formula. The sole purpose of "Please" is to ensure our constant disciplined attention, in the same way as the command, "Attention!" is used to drill soldiers, or the reply, "Thank you so and so" is required of children, as much to train their minds as to teach them good manners.

The experiment continues. The Gardners now hope that one day they will succeed in increasing their pupil's gesture-vocabulary to three or four times this size.

195

What will become of this training when Washoe, at the age of seven or eight, attains sexual maturity? We cannot tell because we know that, as chimpanzees reach puberty, they become more and more independent and increasingly difficult to control.

As far as females are concerned, the matter is quite simple. They quite easily surmount the difficulties of adolescence. As soon as the female chimpanzee has had her first sexual relationship, she generally regains her usual equilibrium. It is a different story with males. Colleagues of mine – Dr Mennerat in Paris and Dr Barraqué in Barcelona – have supplied me with several examples of male chimps they have reared.

Two adult primates, brought up from babyhood by their masters, became so powerful and violent on reaching maturity that they had to be excluded from the family life which they had always shared. There was no alternative but to put them behind bars or send them to zoos. An anthropoid's ability to forget makes this last solution the best way to avoid any dramatic conflict when man, their master, has become in their eyes a rival to be eliminated or a "dominant" to argue with.

As we see, the work of Mr and Mrs Gardner and their predecessors have greatly contributed to a better knowledge of an intelligent anthropoid which has so many points of similarity with man. They have also shown new possibilities of communicating with the animal world.

Other scientists, in their turn, will doubtless follow up the experiments. Other anthropoids will be subjected to similar training, employing either the same techniques or others which are now being tried out.

The "language" which Washoe has learnt cannot,

in any case, be regarded as a one-sided "training". It consists of a common code, by means of which an intelligent animal has established real exchanges with caring human beings.

Only the future can tell whether this "bridge" between monkeys and ourselves, based on affection, is a passing phenomenon, merely temporary, and whether Washoe, once separated from her instructors, will become – like so many of her fellows – a "regressive creature" (to use Kohts's expression), possessing certain human abilities which it has no natural inclination to use or develop; a creature confined within its own narrow circle of inborn imperfections with neither wish nor ability to take the road to progress.

## Notes

[1] Kohts, N.     *Infant Ape and Human Child* (printed in Russian) Museum of Arvinianum (Moscow, 1935)

[2] Hayes, C.     *The Ape in our House*, Gollancz (London, 1952)

[3] Cuny, H.     "Sur la psychologie animale", *Petite Encyclopédie marxiste* (Editions Sociales, 1966)

[4] Gardner, B. and R. Allen     "Un singe à l'Université", *Science et Vie* (December, 1969)

197

# CHAPTER 13

# Conclusion: Homo rex

In the preceding pages I have attempted to sum up the essential facts concerning methods of animal communication, both between animals themselves and in their relationships with man, which animal psychology and neurophysiology have disclosed. From this brief treatment of the subject, one fact emerges: all creatures have their own "language", but, to use J. Filloux's word, they have no "descriptive" language, because none of their languages can express thought. They consist merely of signals relating to each sense (visual, olfactory, tactile and auditory) but these do not stem from a previously established code nor are they consciously applied, as would be necessary for an outsider to understand or reply to them.

Such signals certainly have "informative value"; they produce reactions (in the same way as we respond to indicator lights and sound signals), but neither the animal which gives these signals nor the animal which receives them attaches a symbolic value to them. Today scientists agree that the brain of the higher animals is able to register and associate this information and to transform the resulting actions into an "individual performance". But such actions can never communicate associations of ideas or "thoughts" of a personal nature.[1]

Man is therefore king of his domain. Because of the complexity of his brain (some 14 billion neurons, or nerve-cells, and an unbelievably complicated and active network of synapses – fusions between the cells of the nervous system), only man can, at one and the same time, control the muscles of his mouth cavity (which allows him to articulate a language) and understand the symbolism of that language. Man alone, in the course of his evolution, has attained the different stages of object-maker, artist and reasoning being. But can we state precisely the origins of this symbolic language, which is universally considered to be unique, a biological mark of nobility exclusively reserved for the human race?

## The coming of language

Even the greatest authorities on this subject can only make guesses. Henri Piéron has already posed the question: "Was man's acquisition of language made possible by a form of universal mental mutation, or was it on the other hand, the acquisition of language which brought about such a mutation?" P.-P. Grassé expresses it wittily: "Is man the highest animal or the lowest angel?" This question has never been answered. It is just as difficult to determine, even approximately, at what period in man's history the exceptional privilege of language first appeared.

"Before the written word", writes A. Leroi-

Gourhan, "it is impossible for us to say anything about the use of language and we have little hope of ever finding the flesh of fossilised languages."[2]

Yet Australopithecus (existing approximately one million years ago) probably possessed verbal language. This very ancient ancestor had, in fact, already invented tools, and it is scarcely possible to separate the level of intelligence which could invent a tool from the level of intelligence which could possibly produce "language". Leroi-Gourhan believes that if the language of earliest man went beyond the level of vocal signals, his vocabulary was nevertheless extremely limited.

We can get some idea from studying the language of primitive peoples, some of whom still live today in entirely the same conditions as prehistoric man. The Australian aborigines, the last of the bushmen, the Pygmies, the Papuans of Melanesia and (until a few years ago) a few indomitable Eskimos are classic examples. Although the languages spoken by these primitive peoples vary immensely, even from family to family, a hundred words must have been enough for their ancestors and are still sufficient for their present-day descendants to communicate with each other (abstract ideas or thought excepted).

There is no need to go into further detail about the various theories concerning the origins of human language. Whether speech was present in man from the beginning, the inexplicable privilege of our race alone, or whether it has mysteriously been acquired in the course of man's evolution, will probably always be a matter for debate. But, while taking into consideration that the development of man's skull and brain and the freeing of his hands have no equivalent in any other animal species, we could still perhaps consider the possibility that certain higher mammals might not be incapable of "symbolic thought" of a kind and even of more advanced communication than that of signs.

Experiments at the laboratory of primates in the University of Yale have shown that chimpanzees, taught to insert counters into a slot-machine to obtain sweets, quickly learn to tell the difference between the various colours of the counters corresponding to different types and quantities of sweets. These apes have even been seen stealing counters from one another and exchanging them according to their preferences.

In another area of research into prehistoric man, Professor Dart has pointed out that Australopithecus was already a perfectly organised hunter. This led him to believe that in order to fight successfully against wild cats and rhinos, our ancient ancestors would probably have established an elementary form of co-operation between themselves and that any such co-ordination of action would necessarily imply the use of language. But a pride of lions or a pack of wolves, engaged in similar hunting activities, behave no differently! Each one has his part to play. One animal discovers the position of the prey, another heads it back or cuts off its escape, while another attacks or kills it, but these roles are not fixed. Lions and wolves can change their tactics. Their collaboration varies according to information exchanged between them. In certain well-defined cases, when group co-operation is needed, as in hunting, some of the higher animals may well establish among themselves what Jacques Monod calls a "basic plan".

## The mysterious dog

We are led to wonder whether the dog, which lives in such close contact with man, sharing his immediate environment and giving so many daily examples of spontaneous collaboration, could perhaps have acquired a form of rudimentary animal language which owes nothing to training in the strict sense of the term.

The Gardners, we remember, taught a chimpanzee the rudiments of a language of symbols by means of gestures alone. If these experiments could be pursued, they might add some very encouraging evidence for the first glimmerings of animal thought. If we then think of the domestic dog, it is because, I repeat, this is the only animal which man has never had to subjugate, the only one to have given itself willingly to man, breaking all ties with other animals, to the extent that today it is impossible for it to return to a totally wild existence.

Agreement has not yet been reached on the true origins of the domestic dog. Nevertheless, it is surprising that the majority of the wild dogs (which are relatives if not ancestors of the domestic dog) do not automatically run away at our approach. Lycaons (African carnivores, half hyena and half dog) and dingoes, for instance, watch us from a safe distance. They follow our movements (apparently more from curiosity than from fear), as if demonstrating what a poet might see as "nostalgia for a lost friendship".

On the whole, animals do not run away from man unless experience has taught them aggressive or defensive reactions. On the first expedition to Adélie Land (Antarctica), emperor penguins arrived in their hundreds to look at these strange two-legged creatures who seemed to walk just like them. Suddenly the dogs accompanying the polar explorers rushed at these flightless birds and tore some of them to pieces. In less than a week, the over-confident penguins had learnt the danger represented by man and his dogs.

This brings us to a subtle point. Perhaps for fear of making human comparisons, scientists have never shown so great an interest in studying the psychology of dogs as they have in studying insect behaviour. And yet, what discoveries might be made through a more systematic study of the brains of many different breeds.

Looking at their various shapes and abilities, one is struck by the great physical difference between, for instance, borzoi and pekinese skulls. Can we be certain that their different characters, natural inclinations and obvious neuro-physiological peculiarities are not associated with anatomical differences? For example, it has always been known that a German sheepdog's sense of smell is far better than a bulldog's. But it was only a few years ago that the nasal mucous membranes of the sheepdog were found to contain 200 million ethmoidal cells (relating to smell) while the bulldog's have only half that number. It is not a question of the relative surface area of the mucous membranes: a similar study of man's nasal organs has shown that, irrespective of the length or size of the human nose, it possesses scarcely five million olfactory cells, occupying only a fifth part of the mucous membrane.

A comparative physio-psychological study of the brain of the main breeds of dog has not yet been made. If all dogs, whether long or short, short-headed or long-headed, big or small, have almost identical brains, why are some gifted with an exceptional ability to associate memories and an even more striking intelligence than the great apes? Why do certain dogs, especially sheepdogs, go beyond the level of their training and show "initiative", "judgement" and a sense of "responsibility" which we find quite natural although they know nothing of our moral codes?

We are only just beginning to penetrate this mysterious world of dogs. We have discovered that their minds are frequently and more deeply disturbed than those of other mammals. Canine psychiatry could develop parallel with the study of our own brains. Because dogs share all the pressures experienced by modern man, they too suffer from neuroses and psychoses; they can suffer from loss of will power or nervous debility, become unstable or abnormally aggressive.

It is easy to say that this is merely a question of "animal-neurosis" or "animal-psychosis" because with its four billion or so neurons and its simple cortex, even the most highly developed animal is incapable of abstract "thinking" or "imagining". But what about the agonies suffered by an abandoned pet dog, or the way an epileptic dog will go to its master for protection when it feels a fit coming on? Many more such examples could be listed.

The great apes have not changed since earliest times, nor have the wolf or the coyote. In this matter of evolution, no other animal, tame or wild, can be compared with the attentive dog, which shows such obvious signs of social integration.

We have only to compare the vocal signals between the contented dog and man, with the elementary sounds made by dingo, coyote or wolf, to realise the immense range of expressions which separate the dog of today from what were probably the first domesticated dogs in neolithic times.

It is difficult to attribute this solely to selection and training. Could we also consider the possibility of evolution running parallel with that of man? The domestic dog has been intimately linked with man at all moments during the last ten thousand years, with an apparent rise in status. It was separated from its own kind at the start and excluded from the narrow, communal life it would have had. Since then it has been led gradually to adopt the behaviour of the human beings around it.

All animal societies are unchangeable. Human society, into which the dog has been integrated, has never ceased changing since the end of the paleolithic period. The conditions of the dog's integration are still not fully known, but they are not at all comparable with the social life of wild dogs in packs. Man not only freed the domesticated dog from the need to find its own food, but also defended, protected and freed it from fear of other animals. Either for his own use, or simply for pleasure, man has continuously bred the domestic dog, producing an astonishing range of breeds.

This could perhaps be said of all domesticated animals as well as those under man's control, but it is not altogether true. Keeping to our subject – language – we must stress that only in the highly developed domesticated dog do we have all those different tones of voice, scarcely audible moans, stifled sighs, various tremblings of the body and, above all, those silent looks.

All these signals are so eloquent that Henri Piéron writes quite plainly: ". . . they seem to be expressing an inner language."[3] This might seem rather an exaggerated term to use when we know that even the most intelligent animal, whether dog or chimpanzee, does not have sufficient neurons or brain cells to allow it to think or to communicate "thoughts". Nevertheless, anyone who has owned a happy, contented dog would find it difficult to accept that this domesticated animal is not aware of states of mind. Such awareness – however confused – is something more than mere reflex, and yet less than true "feeling".

Scarcely ten thousand years have passed since the wild dog first became the helper, companion and friend of man. At least five hundred thousand years were needed for Java's Pithecanthropus to develop via Neanderthal and Cro-Magnon man and become *Homo sapiens*. Since then man has entered the electronic age and walked on the surface of the moon.

Rapid though man's development has been since paleolithic times, the development of thought and speech was very slow. It took aeons to achieve the many necessary changes of structure, capacity and formation of the human skull, and the modifications in the development and complexity of the brain.

We know nothing about the dog before the time of *Homo sapiens* and very little about the dog of our own times. But who can say what the dog may one day owe to its unbelievably powerful senses and its wealth of emotions which take the place of "intelligence"?

Brain surgery has shown the link between the intellectual and the emotional halves of an animal's brain. But do we know if intelligence stems from the emotions, or if the emotions themselves give rise to intelligence"?

207

However well equipped we might be to analyse a creature's central nervous system and to understand the functions of the brain (or at least of the parts closest to the surface), it is quite a different matter when we come to analyse an animal's "consciousness". This is why we are often led astray in such research.

The human-like gestures of monkeys, for example, and the relatively long period of maternal care given to their young can mislead us. They have no more to do with real feeling (in this case, awareness of mother-love) than the pose and attitude of the "thinker" adopted by a gorilla or an orang-outang have to do with wisdom and meditation.

The behaviour of dogs can be equally puzzling and can have little or no logical interpretation. Usually, a dog which has never in its life seen a big fire will bark and rush to alert its master the moment it sees such a disaster approaching.

We are content to explain this by saying that flames represent to the dog an abnormal state of affairs, and this makes it react. This is all very well, but imagine a dog which has always lived in the tropics: brought to a new climate, it is suddenly faced with a snow storm, and remains completely indifferent to this phenomenon. Yet this snow is just as strange as the fire. Faced, then, with new experiences, what is merely of interest to a dog, and what is dangerous?

The question leads to another unsolved problem: Do animals remember past experience? "If I could answer this question", writes Lorenz,[4] "I would have found the answer to the problem of body and soul."

This master of the objective school of zoology would like to believe that the higher animals do possess this ability. Unfortunately, although we can easily tell that a dog is sad, we do not know for certain why it is sad. How much harder it is to find out whether the dog knows it is sad!

I hope I will not be criticised for using the dog as my principal witness in the case for animal language. Yet it is probably true that I have spent more time in the company of dogs than with human beings. If the psychology of dogs seems to me more interesting than that of the great apes, it is because I so often feel a communion between us which I cannot explain and which, after fifty years as a veterinary surgeon specialising in dogs, still fascinates me.

Of course, we cannot speak "dog language". The mime, vocalisation and behaviour of dogs are still largely a closed book, but we are aware that they constantly have an irresistible urge to watch us. It is this "exploring look" which makes the dog unique among all animals. It is a sort of bridge which rises up from time to time, as quickly as a mirage, and links us.

Our articulated language is not perceived by a dog as "words" but as intonations. In certain cases, we would be hard put to deny that it attaches a rough symbolic value to these, perhaps using the "simulation function", which is close to imagination. Only the human brain has developed and made full use of this ability, but Jacques Monod[5] believes that higher animals may also be able to use it.

A puppy seeing its master getting ready for a walk demonstrates its pleasure in a familiar way. By anticipation, it simulates all the pleasure that it will, or expects to, enjoy. Through our gestures and slight changes of facial expression, it is mysteriously informed of our current mood and state of mind. This must be because such silent language is more stereo-typed, and standard, more expressive and less deceptive than any words.

To sum up: there is no question of looking forward to the day when the most intelligent and liberated dog (liberated in all senses of the word) will be able to use real language. Our concern is to know whether or not this exceptional companion has bene-fitted mentally from its long journey with man.

"The most complex thought is not necessarily linked with the most complicated parts of the brain.... In more concise terms: the purely materialistic explanation of animal thought seems to be at variance with the facts...."[6] These are not my words. They were written by one of the most experienced animal psychologists, Professor Rémy Chauvin.

## Notes

[1] Monod, J.        *Chance and Necessity*, Collins (London, 1972)
[2] Leroi-Gourhan, A.  *Le geste et la parole*, Albin-Michel (1970)
[3] Piéron, H.        *L'homme, rien que l'homme*, P.U.F. (Paris, 1967)
[4] Lorenz, K.        "Haben Tiere eine subjektives Erleben?", *Münchensamthochschule* (Munich, 1963)
[5] Monod, J.         Op. cit.
[6] Chauvin, R.       *Le comportement animal*, Masson (1969)

# INDEX

Baboon 165, 166, 168–70; behaviour 165; communal life 169; dog-faced 118, 165, 168; "grooming" 166; sexual display 166-7
Badger 119
Basel Zoo 156
Barbary ape, *see* Macaque
Bat 3, 56
Batrachians 57–65 *passim*; hearing 61; mating 62; paroptic vision 58; poisonous secretions 59; self-defence 58; "song" 60; visual display 58
Bear 157
Beaver 121, 155
Bees 7–10, 22; *Bienensprache* ("bee-talk") 22; colour vision 7; "conversation" 8; dance 8; *Rundtanz* 8; *Schwanzeltanz* 8; "scout" 8
Békésy, von 112
Belgium 89, 115
Bernard, Claude 58
*Biologie des lépidoptères* (Paul Portier) 18
Birds 65–102; bird-song 84–91; colonies 82; dancing 75; emotion 79, 80; foveae 66; hearing 83; imitative abilities 88, 100; instinctive behaviour 79; "love-duets" 80; mating 78; migratory 80, 81; "musical language" 85; nocturnal 76; oral communication 65, 68; panoramic vision 66; sea birds 82–3; sexual display 75–6; song-birds 84–91; syrinx 88; talking-birds 91–102; visual stimuli 67, 68
Bison, European 156
Boar 119
Borzoi 204
Bourlière, Prof. 122
Bower bird 76
Brain (mammalian) 109–10; cortex 109, 164
Blackbird 71, 86, 87
Bream 30

Bremond, J. C. 84, 86
Bressou, Prof. Clément 98
Brosset, A. 71, 89
Bronx Zoo, N.Y. 170
Budongo Forest 117
Bulldog 137, 204
Bunting 89
Burnell, Dr. 69
Burton, Maurice 96
Busnel, Dr. R. P. 49, 112
Butenandt, Prof. Adolf 17
Butterfly 14, 15; cataleptic trance 18; Charaxes (genus) 16; coloration 19; defence 19; mating call 17; Portier's discovery 18, 19; ultrasonic waves, sensitivity to 20
Buytendijk 112, 128, 187, 188

Cadéac, Prof. 153
*Calypso* 38
Canary 90
Canary Is. 50, 90
Caudata, *see* Batrachians
Capranica, Prof. 60, 61, 62
Carnivores 71; sensitivity to infra-red rays 3; teleguidance 3
Carp 30
Catfish, African 31
Cat 4, 124, 143–5 *passim*; false lactation 127; mime 144–6; Siamese 125, 144; wild 157, 202
Centre National de l'Acoustique 112
Cetaceans, *see* Marine mammals
Chamois 120, 155
Chapuis, C. 71, 89
Chauchard, Prof. Paul 107, 110, 132
Chauvin, Rémy 4, 12, 13, 21, 22, 79, 100, 210
Chavance, Dr. M.-J. 152
Chimpanzee 164; communication with man 183–97; Goodall, Jane, observations on 171, 172, 173; "Gua" 185 *see also* **Kellogg**; language 191;

Java 93, 207
Jay 76, 94, 96, 99
Joly, R. de 70

Kellogg, Prof. Winthrop N. 46,
 186, 193
Kilimanjaro, Mt. 168
Kiwi 65
Klopfer 153
Kneutgen, Johannes 87
Koehler, Prof. 102
Kohts N. 183–97 *passim*
Kortlandt, Dr. Adrian 173
Kühme, Dr. Wolfdicter 158

Laboratoire d'Ecologie 71
La Cepède 58
La Ferté-Allais 99
Lalitre crab 31
Language, *see* Communication
Lark 71
Lataste, F. 60
Laughter, simian 188
Le Boeuf, B. 115
League for the Protection of
 Birds 99; *see also* Milon
Leroi-Gourhan, Prof. 110, 136,
 200, 201
Lilly, Dr. John 26–50 *passim*,
 73–101 *passim*; predictions 40;
 work with dolphins 40
Lion 157; hunting 202; "lan-
 guage", range of 115, 116
Lobster 25
Lockheed Society 47
Lorenz, Prof. Konrad 4, 72, 89,
 208; Aggression Lecture 134
Los Angeles 34
Lot, Fernand 18
Louis XII 95
Lubbock, John 20
Lycaon 203
Lyre bird 92
Lysergic diethylamin acid 30, 31

Macaque, Japanese 167
Mackerel 30, 32
Magpie 70
Maigres, mating process 28

Mammals 14, 57, 105–13, 132;
 African 116; domesticated
 123–40; European 119; means
 of communication 111, 112;
 "thought" 132; wild 115–22
Mammary glands 108
Manatee 26
Mandrill 168
Mandriota, Frank 31
Mankind *passim*; Australopithe-
 cus 201, 202; brain 200; Cro-
 Magnon man 207; hearing
 61, 71; *Homo sapiens* 207;
 language 199–210; Neander-
 thal man 207; Papuans 201;
 Pithecanthropus 207; pyg-
 mies 201, 207; sense of smell
 155; society 206; speech 22
Marine Experimental Station,
 California 47
Marine mammals 37, 50; brain
 size 37; language 38–46 *pas-
 sim*; reaction to danger 38, 39
Marmot 120, 121
Marsault, Jacques 134
Martens 155
"Martina" 73; *see also* Lorenz
Matthews 75
Max Planck Institute 17, 67,
 87, 158
Meischner 148
Menageries 157, 158
Mermaids, legend of 26
Merwe, Dr. van de 115, 116
Mexico 50, 70
Michel, Aimé 151
Milton, Col. Phillipe 99, 100
Mime 10, 20, 74
Minnow 31
Mocking bird 99
Monod, Jacques 202, 209
Mule 112
Muraena eel 29
Myna bird 93, 94, 98
Myrboy, Dr. Arthur 30

Nassanoff's gland 10
National History Museum,
 London 20

Nerves, sensor-motor 163
Nevada University 191
Newfoundland dog 137
Nightingale 89, 95
Nouvel, Prof. Jacques 99
Norris, Dr. Kenneth S. 46, 49

Oceanography 26, 39
Okapi 116
Olfaction, International Congress on 152
Orang-outang 165, 174, 179, 208
Ornithologist 88, 91
Oscillograph 59
Ostariophysi 30
Ovid 95
Owl 66, 76

Pavlov 33
Pekinese 204
Penguin 76, 77, 78, 203
Peterson, Richard 115
Pfeiffer 31
Phenylethylic alcohol 31
Pheromones 13, 22
Phonemes 21
Piéron, Prof. Henri 23, 146, 200, 206, 207
Pistol-shrimp 29
Poland-Fish, Marie 26
Polar explorers 204
Polarised light, see Bees and Lalitre crab
Porpoise 47–8
Portier, Prof. Paul 18

Quail 76

Rabbit 121, 146
Radioisotopes 4; see also Rémy Chauvin
Rat, white 138; sexual behaviour 154
Réaubourg, M. 51, 53, 55
Regen, Prof. 21
Reynolds, Francis & Vernon 117
Rhinoceros 157, 202

Roach 30
Robin 84
Rostand, Jean 35, 58, 59, 98
Rothschild, Miriam 19
Rundtanz, see Bees

Saint Bernard dog 137
St. Thomas I. 45, 49
St. Thomas's pool 26
Saint-Vincent, Bony de 58
San Diego 47
Spain-Jalouste, Dr. 78
Sardines 32, 34
Schaller, George B. 175, 176, 177
Scheller 20
Schenkel 148, 149
Schleidt, W. & M. 67
Schneider, Dr. Dietrich 17
Scholander, Dr. P. 43
Schwanzeltanz, see Bees
Sea elephant 115
Seagull 4, 70
Sealyham 131
Seamew 34
Seattle Aquarium, see Griffin
Shark 30
Sheepdog 205
Shrimp 28
Shytter (whaling boat) 43
Silène 53
Silkworm moth 17
Siphonia 30
Skunk 155
Solenopsis soevissima 12, 13
South America 93
Spain 89
Sparrow 71, 74, 84
Squirrel 122, 156
Squirrel-monkey 167
Starfish 25
Starling 67, 68
Stimuli, see Communication
Stork 65, 79, 80
Sudan 118
Sumatra I. 179
Sundra Is. 179
Swan 76-7
Szlep 12

216